With my best wishes

Richard E. Putman
June 5th 2001
IPac, 200 - New Orleans

STEAM SURFACE CONDENSERS

BASIC PRINCIPLES, PERFORMANCE MONITORING, AND MAINTENANCE

Richard E. Putman

ASME PRESS
NEW YORK

Copyright © 2001
The American Society of Mechanical Engineers
Three Park Avenue, New York, NY 10016

Library of Congress Cataloging-in-Publication Data

Putman, Richard E., 1925 –
   Steam surface condensers: basic principles, performance monitoring, and
   maintenance / Richard E. Putman.
      p.   cm.
   Includes index.
   ISBN 0-7918-0151-9
   1. Condensers (Steam)   I. Title.
   TJ557 .P88 2001
   621.1'97—dc21   00-064607

Statement from By-Laws: *The Society shall not be responsible for state-ments or
opinions advanced in papers . . . or printed in its publications* (B7.1.3)

INFORMATION CONTAINED IN THIS WORK HAS BEEN OBTAINED BY
THE AMERICAN SOCIETY OF MECHANICAL ENGINEERS FROM SOURCES
BELIEVED TO BE RELIABLE. HOWEVER, NEITHER ASME NOR ITS
AUTHORS OR EDITORS GUARANTEE THE ACCURACY OR COMPLETENESS
OF ANY INFORMATION PUBLISHED IN THIS WORK. NEITHER ASME NOR
ITS AUTHORS AND EDITORS SHALL BE RESPONSIBLE FOR ANY ERRORS,
OMISSIONS, OR DAMAGES ARISING OUT OF THE USE OF THIS INFOR-
MATION. THE WORK IS PUBLISHED WITH THE UNDERSTANDING THAT
ASME AND ITS AUTHORS AND EDITORS ARE SUPPLYING INFORMATION
BUT ARE NOT ATTEMPTING TO RENDER ENGINEERING OR OTHER PRO-
FESSIONAL SERVICES. IF SUCH ENGINEERING OR PROFESSIONAL SER-
VICES ARE REQUIRED, THE ASSISTANCE OF AN APPROPRIATE PROFES-
SIONAL SHOULD BE SOUGHT.

For authorization to photocopy material for internal or personal use under circumstances not falling within the fair use provisions of the Copyright Act, contact
the Copyright Clearance Center (CCC), 222 Rosewood Drive, Danvers, MA 01923,
Tel: 978-750-8400, www.copyright.com.

To Agnes
Wife and Mother to Caroline and William
Best Friend and Constant Companion
for nearly 40 years
and a
Fellow–Student in the University of Life

TABLE OF CONTENTS

Chapter 5   Interactive Model of Condenser and
Low-Pressure Turbine Stage   77

Chapter 6   Case Studies   91

Chapter 12　Air and Water Inleakage Detection and Eddy-Current Testing　195

Chapter 13   Performance Monitoring of Power-Plant Heat Exchangers   225

Appendix A   Mathematical Procedures   239

INTRODUCTION

None of us, whether in our personal or professional capacity, is an island. With almost no exceptions, our attempts to extend or consolidate the technology we know derive from the work of numerous predecessors. As part of an evolving process, we first draw on the past to become familiar with the extent of existing knowledge, for only then can we proceed further. In a letter to Robert Hooke written in 1675, even Sir Isaac Newton confessed that he had stood "on the shoulders of Giants"; if Newton was gracious enough to acknowledge his debt to the past, so, in our own small way, must we.

This book has, of necessity, drawn on the ideas of several giants. Some are famous: Fibonacci, Kern, Newton, Nusselt, Raphson, Rankine, and Silver are all referenced. But many ideas have been drawn from that multitude of diligent practitioners, known and unknown, who, each in a unique way, have contributed to the body of knowledge we now enjoy and continue to build on today.

The information presented in this book is the result of a program of investigation and research into the behavior of power-plant condensers, extending over several years. The project began as a quality assurance program, intended to compare results obtained from routine condenser-tube heat transfer tests with expected values calculated according to standard procedures of both the Heat Exchange Institute (HEI) and the American Society of Mechanical Engineers (ASME). The differences between the two procedures soon became noticeable, and the ASME method, being in closer accord with modern heat transfer theory while also able to take both shell- and tube-side conditions into account, was subsequently chosen as the quality control standard for such tests. Not only did the use of the ASME values permit the test rig to be calibrated: they also allowed the condition of tube samples received from the field to be quantified, and the results of cleaning tubes using a number of different methods to be compared with confidence.

By the early 1990s, the utility industry had come to recognize that condenser performance within the Rankine cycle was a significant factor in improving unit heat rate and/or increasing generator capacity. As a result, I received frequent requests to analyze the behavior of operating condensers, with a view to improving it. The traditional performance criterion was to compare the current cleanliness factor with the design value, but deviations in this factor were difficult to express in terms of increased cost. To meet this industry need, we set out to develop a technique for modeling condenser performance so as to quantify avoidable heat losses due to fouling as well as computing the degree of fouling, or fouling factor. We also wanted the technique to model the various equipment configurations encountered in the field without major reprogramming. Condenser configurations included one-, two-, or three-compartment condensers, each compartment having one or two passes; the details of the water flow path connecting the compartments also had to be reflected. Because we had already established the reliability of the ASME method in predicting single-tube heat transfer coefficients, we chose it as the basis of the heat transfer equations to include in these models.

However, it soon became obvious that modeling the condenser alone was not enough, because of the natural interaction between condenser behavior and subsequent response from the steam turbine governor. Turbine thermal-kit data was carefully analyzed and models of the expansion line were generated, designed to reflect the relationships between the enthalpy and flow of the exhaust vapor with respect to back pressure as well as with respect to generated power. In the course of this work, we found that the models should include the annulus exhaust losses and also respond to the occurrence of annulus choking, and we included these features.

A knowledge of current cooling-water flow rate is vital if heat transfer coefficients and thermal resistances are to be compared in a meaningful way, but few plants are provided with the means to measure such flows. In due course, we found that these models could be extended so that condenser duty could be used as the basis for estimating cooling-water flow rate. This work proved to be very successful and has been outlined in the case studies in Chapter 6.

Our success in developing this set of interactive condenser performance models allowed us to establish the degree of fouling and to quantify the avoidable heat loss (i.e., the difference between current condenser duty and the duty if the condenser were clean) and to express it as a monetary saving. However, this data was calculated from snapshots of condenser operating data; while establishing the momentary state of a condenser is useful, the *rates* at which fouling and air ingress have developed are equally important factors in plant maintenance decision making.

Cleaning a condenser can be quite costly, not so much on account of the cost of the cleaning operation itself, but mainly because of the power that either is not generated or has to be purchased while the cleaning is in progress. Optimal maintenance decision making also requires a historical perspective, and the data we uncovered gave us that perspective. We explored several possible methods for establishing an optimal cleaning frequency and determining in which months the cleaning should be performed. In the course of this activity, we became aware of the need for a fouling model for the condenser. Compartmental fouling resistances could already be calculated at sequential points in time; after analysis, these values came to form the basis for the necessary models and would also become a useful means for studying the reason for changes in the rate of fouling or whether air leaks had developed. Further, since the behavior of a given condenser, when clean, can be characterized by the cooling-water inlet temperature and turbogenerator load, it became apparent that the mean monthly average water temperatures and load (or capacity factor) would be needed. All this led to the evaluation of various cleaning-schedule optimization techniques based on this set of historical data, and these have proven to be very helpful to plant maintenance departments.

As a result of electricity deregulation, many power plants are no longer base-loaded but cycle through their load ranges once or twice per day. In the past, condenser performance monitoring has focused on full-load conditions, so that current performance could easily be compared with original design data. However, although the HEI design cleanliness factor has been the traditional parameter for judging the state of a condenser, we found that this was not a constant as had usually been assumed, but could demonstrably be shown to vary with load. The relationship between clean-condenser cleanliness factor and generated power was then added to the models so that fouling losses and associated resistances could be estimated with greater accuracy.

The audience for this book includes not only practicing power-plant engineers but also marine engineers and students of heat transfer equipment, as well as engineering instructors seeking examples of practical engineering problems whose analysis and solution employ science and mathematics in some interesting ways. The early chapters supply background information when needed. Chapter 1 reviews basic heat transfer principles involved in monitoring condenser performance. Chapter 2 examines past practice in this field, offers a critique of the method adopted by the Heat Exchange Institute (HEI), outlines the thermal resistance method used in various ASME standards, and provides comparative calculations based on these two standards. Chapter 3 identifies some of the objectives to be followed in developing a set of condenser performance models and examines the instrumentation required for performance monitoring together with its application within a plant.

Chapter 4 focuses on the analysis of the turbine low-pressure expansion line and the process for generating mathematical models of enthalpy versus back pressure for various loads. These can be used to estimate the cooling-water flow rate, but they are also an essential part of the interactive condenser/low-pressure turbine performance models.

Chapter 5 reviews the structure of the basic interactive condenser/low-pressure turbine model, the set of boundary conditions and constraints, and the mathematics required to solve a set of nonlinear simultaneous equations. The chapter also describes how to expand the basic model structure to include more than one condenser compartment.

Chapter 6 includes two case studies. The first examines the calculation of the cooling-water flow rate for a number of plants. The second is for a tidewater plant and concentrates on the correlation between the fouling, water flow rate, and avoidable loss calculations.

Chapter 7 outlines the various factors involved in developing an optimum condenser cleaning schedule and different ways of approaching the return-on-investment criteria. Development of the database from unit operating data and condenser performance calculations is also outlined.

Chapter 8 examines the economic impact of condenser performance on the cost of power-plant operations. A method for converting avoidable losses to their equivalent carbon emissions is also included, in the event that the carbon dioxide provisions of the Kyoto Treaty should come into effect.

Chapters 9 through 12 deal with issues of diagnosing and remedying condenser fouling. Chapter 9 approaches the diagnosis of changes in condenser behavior, lists some typical malfunctions, and examines some diagnostic methods that have been employed. Chapter 10 examines the various types of fouling and/or corrosion which can occur and discusses some common types in detail. In particular, the galvanic corrosion of manganese with stainless steel is discussed; while the monitoring of copper concentrations in condenser cooling-water effluent is also examined. Chapter 11 is concerned with methods for cleaning condenser tubes in situ. The chapter compares various cleaning tools and methods and also reviews problems of personnel safety when working in enclosed spaces such as condenser waterboxes. Chapter 12 is devoted to a discussion on the detection and location of both cooling-water and air inleakage into the condenser shell, as well as their effects on the condenser and unit performance. Chapter 12 also describes the use of eddy-current testing to locate and estimate the depth of pitting or holes in tube walls caused by corrosion or erosion mechanisms.

Chapter 13 offers an introduction to the performance of power-plant heat exchangers. For more information, readers can refer to the TEMA standards that are the principal source of this information.

Finally, included in Appendix A are the principles on which the various mathematical solutions are based, especially various ways of conducting numerical analysis of the data and solving equations of various types, regression analysis of operating data, and the Newton-Raphson method for solving sets of nonlinear simultaneous equations. The brief Appendix B displays FORTRAN routines for determining properties of saline water.

ACKNOWLEDGMENTS

I would like to acknowledge with sincere thanks the support of this research by Conco Systems, Inc. of Verona, Pennsylvania, which has extended over many years. Special thanks are also due to Mr. George E. Saxon, Sr., and Mr. Edward G. Saxon, Chairman and President of the company, respectively, for encouraging me in pursuing the project; and to Mr. George E. Saxon, Jr., for his invaluable help in developing the scope of the book, providing both data and background information, and reviewing the manuscript at various stages of its development. Thanks are also due to the coauthors who collaborated with me on the many papers cited, and to colleagues who provided the operating data for analysis from which many of the insights were drawn. The book has been greatly improved by suggestions made by Tom Rabas and Don McCue, to both of whom I extend my sincere appreciation. Finally, I would like to express my deep thanks to my wife, Agnes, for her active support, especially in being willing to give up so much of the time we might otherwise have spent together doing other things, in order that this book should be written.

NOMENCLATURE

A	Tube outside surface area	ft^2
C_p	Specific heat of water at bulk temperature	$\text{Btu}/(\text{lb}_\text{m} \cdot {}^\circ\text{F})$
d_i	Inside diameter of tube	in
d_o	Outside diameter of tube	in
D_o	Outside diameter of tube	ft
g_c	Acceleration due to gravity	$= 417 \times 10^6 (\text{ft} \cdot \text{lb}_\text{m})/(\text{h}^2 \cdot \text{lb}_\text{f})$
G	Cooling-water flow	GPM
h	Heat transfer coefficient	$\text{Btu}/(\text{ft}^2 \cdot \text{h} \cdot {}^\circ\text{F})$
h_f	Steam-side heat transfer coefficient	$\text{Btu}/(\text{ft}^2 \cdot \text{h} \cdot {}^\circ\text{F})$
k	Thermal conductivity of water film	$\text{Btu}/(\text{h} \cdot \text{ft} \cdot {}^\circ\text{F})$
k_f	Thermal conductivity of conductive film	$\text{Btu}/(\text{h} \cdot \text{ft} \cdot {}^\circ\text{F})$
k_m	Thermal conductivity of metal	$\text{Btu}/(\text{h} \cdot \text{ft} \cdot {}^\circ\text{F})$
L	Tube length	ft
LMTD	Log mean temperature difference	${}^\circ\text{F}$
m	Rate of condensation per unit area	$\text{lb}_\text{m}/(\text{h} \cdot \text{ft}^2)$
N	Number of tubes	
N_pass	Number of passes	
Nu	Nusselt number	
q	Heat flux	Btu/h
Pr	Prandtl number	
R_f	Steam-side resistance	${}^\circ\text{F}/(\text{Btu}/\text{h} \cdot \text{ft}^2)$
R_foul	Fouling resistance	${}^\circ\text{F}/(\text{Btu}/\text{h} \cdot \text{ft}^2)$
R_t	Tube-side resistance	${}^\circ\text{F}/(\text{Btu}/\text{h} \cdot \text{ft}^2)$
R_w	Wall resistance	${}^\circ\text{F}/(\text{Btu}/\text{h} \cdot \text{ft}^2)$
Re	Reynolds number	
T_b	Bulk water temperature	${}^\circ\text{F}$
T_c	Condensate film temperature	${}^\circ\text{F}$
T_in	Cooling water inlet temperature	${}^\circ\text{F}$
T_out	Cooling water outlet temperature	${}^\circ\text{F}$
T_v	Vapor temperature in shell	${}^\circ\text{F}$
T_{wo}	Temperature of tube outer surface	${}^\circ\text{F}$
U	Overall heat transfer coefficient	$\text{Btu}/(\text{ft}^2 \cdot \text{h} \cdot {}^\circ\text{F})$

V	Water velocity	ft/s
W_{EX}	Exhaust flow per tube foot	$lb_m/(h \cdot ft)$
W	Mass flow	lb_m/h
λ	Latent heat of condensation	Btu/lb_m
μ	Viscosity of water at bulk temperature	$lb_m/(h \cdot ft)$
μ_f	Viscosity of condensate film	$lb_m/(h \cdot ft)$
ρ	Liquid density	lb_m/ft^3

Chapter 1

BASIC HEAT TRANSFER PRINCIPLES

Steam surface condensers are an important component in every steam turbogenerator unit operating in the electric utility industry. They have a threefold purpose: (1) to maximize the electric power conversion of the energy of the steam supplied to the turbine throttle; (2) to recover vapor exhausted from the low-pressure stage of the turbine in the form of condensate, which can then be recycled directly to the boiler; and (3) to recover the heat in the condensate and so reduce the amount of heat required of the boiler to raise the steam to the desired throttle pressure and temperature conditions. Almost all steam turbogenerator units, whether nuclear or fossil-fired, operate in accordance with the Rankine cycle.

Combined-cycle plants operate in accordance with the Brayton cycle, and include heat recovery steam generators (HRSGs) in which steam is generated from the heat contained in the exhaust from the gas turbine(s). This steam is then passed to a steam turbine, thus maximizing the amount of power generated from the fuel consumed by the gas turbines. Steam turbines operating within the Brayton cycle are also equipped with condensers, which perform the same functions as those in turbines operating within the Rankine cycle. In fact, the steam side of the HRSG, the steam turbogenerator, the condenser, and the feedwater heaters together constitute a Rankine cycle subsystem.

In the process industries, it is quite common for steam to be raised in boilers at high efficiency, either directly from fossil fuels or, as in the pulp and paper industry, from recycled by-product fuels, and then to be passed to cogenerating steam turbogenerators. In this way, power is generated by the process steam as it expands through and/or is extracted from the turbogenerators. Most of these steam turbogenerators are also equipped with condensers, which allow as much condensing power as is needed to be generated from the heat supplied to the boilers, while also fulfilling the other functions of condensers included within the Rankine cycle. Cogenerating steam turbogenerators equipped with condensers also free the power generation from a rigid dependency on the amount of process steam being consumed. This is in contrast to those machines in which the flow of steam to the throttle is controlled by the back pressure. In these cases, the amount of power generated is a direct function of the amount of process steam which passes through, and the generated power and process steam flow are closely coupled.

Many countries in the Middle East also use condensing steam turbogenerators to generate potable water.

In addition to their role in condensing the vapor exhausted from the low-pressure stage of steam turbogenerators, the principles of condensation are also used within the Rankine cycle in the design of feedwater heaters, which use steam extracted from intermediate stages of the turbogenerator to preheat the feedwater before it enters the boiler. However, while the turbine exhaust vapor conditions often lie within the wet zone, the steam supplied to feedwater heaters at higher pressures is often extracted under superheated conditions.

Steam surface condensers induce condensation of vapor by removing its latent heat, using cooling water drawn from rivers, lakes, or the seashore, or water circulating within a closed system equipped with cooling towers. Clearly, the heat removed by this water must be discharged into the environment. Thus, in addition to conserving the mass and energy circulating within the power generation system, condensers have the added burden of minimizing these heat discharges, not only because of their direct thermal impact on the environment, but also because of the additional emissions of carbon dioxide and other noxious gases associated with burning the fuel equivalent of any avoidable heat of condensation discharged.

In economic terms, the maintenance of power-plant steam surface condensers can have an important impact on unit heat rate. In nuclear plants, or in fossil plants operating under conditions of high circulating water temperature, proper condenser maintenance can also ensure that generation capacity is not curtailed. Thus, the fundamentals of condensation and the associated heat transfer mechanisms, the monitoring of condenser performance, and proper condenser maintenance are all important considerations in the successful management of steam power plants.

1.1 SINGLE-TUBE HEAT TRANSFER RESISTANCES

The purpose of the condenser is to remove latent heat from the exhaust vapor from the low-pressure stage of the turbine. Tubes are placed in the path of the vapor, and as cooling water is circulated through them, the vapor condenses on their cool outer surface, so that its temperature drops from the vapor temperature T_v to the tube surface temperature T_s (see Figure 1.1). However, the latent heat can be removed only if it is allowed to leave the vapor space by passing through the condensate film. In fact, modern heat transfer theory assumes the presence of four major films or layers through which this same quantity of heat must flow in series, namely:

- Condensate film
- Tube wall
- Fouling layer (if any)
- Water-side boundary film

Each of these films or layers has its own thermal resistance with units $°F/(Btu/(ft^2 \cdot h))$, some of these resistances varying with temperature level, temperature differences, and even flow conditions. By summing these thermal resis-

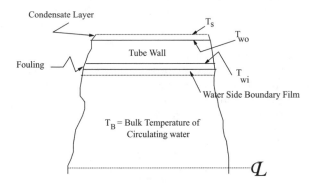

FIGURE 1.1. Tube thermal resistances.

tances, the total thermal resistance between the vapor space and the circulating water can be estimated. The reciprocal of this resistance sum is the overall heat transfer coefficient of the tube, with units Btu/(ft^2 · h · °F) and usually expressed with reference to the outer tube surface.

In discussions of single-tube heat transfer resistances, the calculations are normally associated with a single tube located in the first (upper) row of the condenser tube bundle which is encountered by the vapor as it leaves the turbine exhaust. Other tubes in the bundle are subjected to different operating conditions, and their thermal resistances can also be calculated if the operating conditions are known. However, it will be shown that there are several ways of relating the single-tube heat transfer resistance to overall condenser performance; in this way, a need for detailed knowledge of the distribution of operating conditions throughout the condenser can be avoided.

1.1.1 Condensate Film

Hewitt et al. [1994] identify three types of vapor condensation which can occur in steam surface condensers. *Homogeneous condensation* can take place under either an increase in pressure or a decrease in temperature of the vapor. It commonly occurs in the low-pressure stages of a turbine, leading to the formation of entrained condensate droplets. Although these droplets are normally too small to cause severe problems, their deposition and reentrainment can sometimes lead to blade erosion.

Dropwise condensation can occur when the outer tube surface has poor wetting characteristics. Dropwise condensation has a relatively high heat transfer coefficient but rarely occurs in a steam surface condenser. Tanasawa [1989] has reviewed the numerous papers on dropwise condensation published in the *Transactions of the Japan Society of Mechanical Engineers;* while Qi Zhao [1994] of

Heriot Watt University has published several reports on methods of treating the outer surfaces of tubes to promote dropwise condensation. Finally, Rifert et al. [1989] of Kiev Polytechnical Institute have shown how the use of a fluorine-containing disulfide in the condensers of desalination plants can promote dropwise condensation, any residual fluorine in the potable condensate presumably enhancing dental health in the local community.

Filmwise condensation is the form most frequently encountered in condensers and feedwater heaters in the utility industry. Nusselt, in a classic paper [1916], concluded that the heat transfer coefficient through the condensate film on the shell side of a horizontal tube is given by:

$$h_f = 0.725 \left[\frac{k_f^3 \rho^2 g \lambda}{\mu_f D_o (\Delta T)} \right]^{0.25} \tag{1.1}$$

where the mean temperature of the condensate, used to calculate its physical properties, is given by:

$$T_c = \frac{T_v + T_{wo}}{2} \tag{1.2}$$

Note that Nusselt's value of T_c was calculated directly from the given data, and he claimed that any error in calculating the film properties by using this value is negligible. Nusselt also defined the value of ΔT as:

$$\Delta T = T_v - T_{wo} \tag{1.3}$$

but he did not specifically address the actual value of tube outer surface temperature T_{wo}. A method for rationally estimating its value will be discussed later.

Although it may have been originally suggested by others, Silver [1963–1964], in another classic paper, shows that the heat transfer coefficient, h_f, can be expressed alternatively by the following important relationship:

$$h_f \Delta T = m\lambda \tag{1.4}$$

Substituting in Equation (1.1) and then solving for h_f yields the following relationship:

$$h_f = 0.6513 k_f \left(\frac{\rho^2 g}{\mu D_o m} \right)^{0.333} \tag{1.5}$$

Now, $W_{EX} = \pi D_o m$. Substituting in Equation (1.5) and rearranging gives:

Table 1.1. Regression Coefficients for Condensate Properties

Property	Intercept	T	T**2	T**3
Specific heat	1.0244	-7.1715E-04	6.1796E-06	-1.6397E-08
Density	62.087	2.2519E-02	-3.3873E-04	1.0579E-06
Viscosity	3.2211	-4.4011E-02	2.2023E-04	-3.3678E-07
Conductivity	0.30488	8.4393E-04	-1.2922E-06	-6.1078E-10

$$h_f = 0.954 k_f \left(\frac{\rho^2 g}{\mu W} \right)^{0.333} \tag{1.6}$$

Equation (1.1) is used to calculate the single-tube film heat transfer coefficient, while Equation (1.6) is used when a single tube is placed in the path of the exhaust flow from the low-pressure stage of a turbine. Both Equations (1.1) and (1.6) are included in Appendix 2 of ASME PTC.12.2-1983 [ASME] and state the conductance of the condensate film in the same units as heat transfer coefficients, namely, Btu/(ft^2 · h · °F), while the resistance of the condensate film is the reciprocal of its conductance. This Appendix also states that the physical properties are to be evaluated at condensate film temperature T_{wo} and displays the properties in the form of curves. These have been regressed as third-order polynomials having the sets of coefficients shown in Table 1.1. Note that the coefficients for viscosity in the table calculate its value in centipoise as a function of temperature. To convert to viscosity in lb$_m$/(h · ft), as required by Equations (1.1) and (1.6), this viscosity in centipoise must be multiplied by 2.42.

All the foregoing comments apply to steam or vapor with conditions lying on or below the saturation line, as defined in the ASME Steam Tables [ASME 1993].

Superheated Steam

The condensation of superheated steam is slightly more complicated in that the temperature of the superheated vapor (T_G) at the tube outside surface must first be reduced to below its saturation value before condensation can commence. Hewitt et al. [1994] show that the adjusted film conductance for superheated steam can be obtained from:

$$h_{\mathrm{corr}} = h_f (1 - \xi)^{0.25} \tag{1.7}$$

where

$$\xi = \frac{c_p G (T_G - T_{\mathrm{sat}})}{L_{LG}} \tag{1.8}$$

Other Condensate Heat Transfer Issues

Since air and other non-condensibles can enter the vapor space of the condenser, by being either dissolved in the makeup water or generated by the chemicals used for feedwater treatment, or through air inleakage, air-ejection apparatus is normally provided to remove a designated amount of these non-condensible gases. However, leaks can develop which allow atmospheric air to ingress directly into the vapor space. These non-condensible gases, regardless of their source, cause the condenser tube surfaces to become blanketed by a mixture of air and vapor, which tends to lower the heat transfer coefficient at the condensate film interface. Non-condensible gases are also commonly present in geothermal steam used for power generation; this topic will be examined in more detail later in this chapter.

Another condensate issue is that of subcooling. Clearly, optimum operation of a steam turbine condensing system requires that only the latent heat of condensation be removed and that the condensate temperature in the hot well not be allowed to fall below the corresponding vapor saturation temperature value. This topic will also be discussed further in Chapter 2, on condenser performance monitoring.

1.1.2   Tube Wall Resistance

Kern [1990], along with numerous others, presents a derivation of the thermal resistance for a tube wall, with material of thermal conductivity k_m, in the following way. Consider a 1-ft length of tube, with the source of heat on the outside, and consider an elemental ring of radius r and thickness dr having a temperature difference across the ring of dt. Then the heat flux flowing into the ring can be given by:

$$q = 2\pi r k_m \frac{dt}{dr} \tag{1.9}$$

which, after rearranging and integrating, gives:

$$t = \frac{q}{2\pi k_m} \log r + C \tag{1.10}$$

Now, when $r = r_o$, $t = t_o$; and when $r = r_i$, $t = t_i$. Thus:

$$t_i - t_o = \frac{q}{2\pi k_m} \ln\left(\frac{r_o}{r_i}\right) \tag{1.11}$$

Converting to thermal resistance R_w, substituting diameters for radii, and adjusting to provide resistance with respect to the outer surface of the tube ($\pi d_o/12$) gives the following:

$$R_w = \frac{d_o}{24k_m} \ln\left(\frac{d_o}{d_i}\right) \qquad (1.12)$$

Values for thermal conductivity k_m included in Equations (1.11 and 1.12) are listed in the TEMA Standard [1988] for a wide range of tube materials.

1.1.3 Fouling Resistance

Fouling deposits normally occur on the inside surfaces of tubes, although on rare occasions they have been found on outer tube surfaces-in those cases, because of faulty boiler feedwater treatment. Some of the causes of fouling, as well as methods for its removal, will be discussed in detail in Chapters 8 through 11. The presence of a fouling film presents an additional thermal resistance to heat flux from the vapor to the cooling water flowing on the inside of the tubes and is designated R_{foul}. Note that the presence of air in the vapor space has a similar negative effect on heat transfer to that of fouling. Thus a falling off of condenser performance may not be due to fouling alone; and care should be taken in the diagnostic procedure to distinguish between these two possible causes. Note also that when fouling of the outer tube surfaces occurs, it is not independently identifiable, since it has the same effect as inner-wall fouling. Tube heat transfer tests are required to quantify these two separate effects.

1.1.4 Water Film Resistance

To calculate the tube-side water resistance, the Rabas/Cane correlation [1983] is used:

$$\mathrm{Nu} = 0.0158 \mathrm{Re}^{0.835} \mathrm{Pr}^{0.462} \qquad (1.13)$$

If

$$\mathrm{Re} = \left(\frac{\rho d_i V}{\mu}\right) \quad \mathrm{Pr} = \left(\frac{C_p \mu}{k}\right) \quad \text{and} \quad \mathrm{Nu} = \left(\frac{h_f d_i}{k}\right)$$

then, converting from the heat transfer coefficient of Equation (1.13) to film thermal resistance:

$$R_t = 0.045057 \left(\frac{\mu^{0.373}}{\rho^{0.835} k^{0.538} C_p^{0.462}}\right)\left(\frac{d^{0.165}}{V^{0.835}}\right)\left(\frac{d_o}{d_i}\right) \qquad (1.14)$$

Note that this correlation uses the properties of water at *bulk* temperature T_b, defined as the mean of the inlet and outlet cooling-water temperatures. Neither the temperature at the inner tube surface nor that at the fouling deposit surface is explicitly taken into account. Furthermore, the density and specific heat should be calculated as a function of the salinity of the water (see Figures 1a and 1b of PTC.12.2 [ASME 1983]), while conductivity and viscosity can be calculated from the regression coefficients of Table 1.1, which are based on Figures A-1 and A-3 of PTC.12.2 (see also Appendix B). Note that the term (d_o/d_i) at the right-hand end of Equation (1.14) allows the water film resistance to be referred to the outer surface of the tube.

1.1.5 Single-Tube Heat Transfer Coefficient

Having calculated the values of the various thermal resistances, we can obtain the overall heat transfer coefficient from the following, where R_{foul} is the fouling resistance:

$$U = \frac{1.0}{1/h_f + R_w + R_{foul} + R_t} \tag{1.15}$$

1.1.6 Calculation of Condensate Temperature at Tube Surface

It was remarked earlier that, while the value of the condensate film heat transfer coefficient h_f was dependent on the condensate film temperature, its actual value was uncertain. To obtain an estimate of this temperature, consider the following relationship between thermal resistances and temperature differences:

$$\frac{T_{wo} - T_b}{T_v - T_b} = \frac{R_w + R_f + R_t}{1/h_f + R_w + R_f + R_t} \tag{1.16}$$

Clearly, with clean tubes, R_f is zero, and the relationship must be evaluated accordingly. The solution for the value of T_{wo} is an iterative method and involves making a first estimate, calculating the corresponding shell-side resistance and substituting it into Equation (1.16). In decrementing the estimate for T_{wo}, recalculating the steam-side thermal resistance, and reevaluating Equation (1.16), there will be a point when the difference between the left- and right-hand sides of Equation (1.16) changes signs. This provides a close approximation to the solution. An even closer approximation can be obtained by using a more sophisticated numerical method such as Newton-Raphson or a Fibonacci search.

1.2 TUBE HEAT TRANSFER AS A FUNCTION OF OPERATING CONDITIONS

Following Kern [1990], consider a tube mounted in the vapor space of a condenser, with vapor temperature T_v, mass cooling-water flow rate W, cooling-water inlet temperature T_{in}, and cooling-water outlet temperature T_{out}, together with a constant (mean) value of the heat transfer coefficient U averaged along the length of the tube. The heat flux dQ through any elemental tube disk of surface area dA, having a temperature difference across it of Δt, is given by:

$$dQ = U \, dA \, dT \tag{1.17}$$

Now, from the slope of the straight line relating Q to ΔT,

$$\frac{d\Delta T}{dQ} = \frac{\Delta T_2 - \Delta T_1}{Q} \tag{1.18}$$

Rearranging Equations (1.18) and (1.19) and eliminating dQ gives:

$$\frac{U}{Q}(\Delta T_2 - \Delta T_1)\int_0^A dA = \int_{\Delta T_1}^{\Delta T_2} \frac{d\Delta T_1}{\Delta T_1} \tag{1.19}$$

Integrating this expression gives:

$$Q = \frac{UA(\Delta T_2 - \Delta T_1)}{\ln\left(\dfrac{\Delta T_2}{\Delta T_1}\right)} \tag{1.20}$$

or, in terms of water and vapor temperatures,

$$Q = \frac{UA(T_{out} - T_{in})}{\ln\left(\dfrac{T_v - T_{in}}{T_v - T_{out}}\right)} \tag{1.21}$$

or

$$\frac{Q}{UA} = \frac{T_{out} - T_{in}}{\ln\left(\dfrac{T_v - T_{in}}{T_v - T_{out}}\right)} \tag{1.22}$$

The right-hand side of Equation (1.22) is termed the *log mean temperature difference (LMTD)*, given the unique values of A and U. Equation (1.21) is the conventional overall heat transfer equation for a condenser, in which the water temperature rise is assumed to be linear and the heat transfer coefficient U is assumed to be constant along the length of the tube (Figure 1.2). While Silver [1963–1964] termed Equation (1.21) the Grashof equation, both Hewitt et al. [1994] and Kern [1990] termed it the Fourier equation, Kern giving his reasoning. Equation (1.21) will therefore be termed the *Fourier equation* in the remainder of this text.

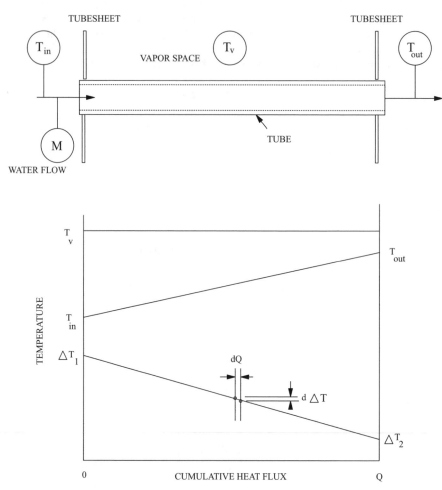

FIGURE 1.2. Heat transfer through tube as a function of operating conditions.

1.3 TEMPERATURE PROFILES ALONG CONDENSER TUBES

The basic heat transfer equation (1.21) was derived assuming a constant value of the U coefficient and a linear relationship between heat flux and distance along the tube length. However, both Hewitt [1994] and Silver [1963–1964] suggest that these assumptions are not literally true. To explore these relationships in greater depth, we examined the behavior of the condenser on a 500-MW unit, assuming a cooling-water flow rate of 250,000 GPM, equivalent to a tube velocity of 7 fps, an inlet water temperature of 70°F, and a shell vapor temperature of 95.51°F, equiv-

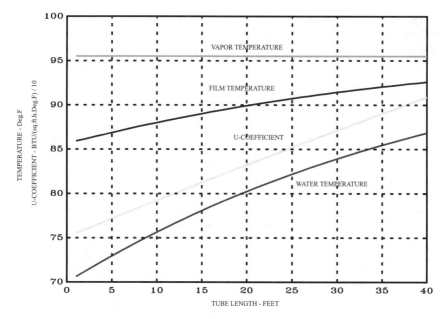

FIGURE 1.3. Temperature profiles—500 MW, 70°F, 250,000 GPM.

alent to a back pressure of 1.7 in-Hg. The tubes in this condenser were 40 ft long and 0.875 in outside diameter and were made from 18 BWG admiralty metal. To carry out the analysis, we calculated the water-film and tube resistances, together with the steam-side resistance, for the first 1-ft length of a clean tube and computed the heat transfer coefficient for this elemental 1-ft length using Equation (1.15). In the course of calculating the shell-side resistance, we also computed the temperature of the condensate film at the outer tube surface T_{wo}.

The heat flux through this 1-ft length of tube element i can be calculated from:

$$Q_i = U_i A_i (T_v - T_{in})$$

Knowing the mass water flow rate through the tube, water density, specific heat, and inlet temperature, we can now compute the temperature of the water leaving this 1-ft section i. This becomes the inlet water temperature to the succeeding 1-ft section $i + 1$, and the calculations are repeated for all 40 feet. The temperature of the water leaving the 40th length element is the tube water outlet temperature T_{out}.

Figure 1.3 shows a plot of the results for this case. The heat transfer coefficient increases as the water temperature rises along the length of the tube. However, if the appropriate values of T_v, T_{in}, T_{out}, tube surface area A, and the heat acquired by the cooling water Q are substituted in Equation (1.21) and the relationship is then solved for U, its value will be very close to the mean of the heat transfer coefficients calculated for each 1-ft length. Thus the value of U incorporated in Equation (1.21) is not a constant but is the mean value of the coefficients calculated for the individual elements summed over the whole tube length.

The plot of condensate film temperature at the tube outer surface, as shown in Figure 1.3, also deserves comment. Silver and others have indicated that the film temperature T_{wo} must, clearly, be lower than the vapor temperature T_v if any heat is to flow through the condensate film and on through the tube wall. Figure 1.3 shows that the difference in temperature can be significant, in this case varying from almost 10°F at the tube inlet to about 2°F at the tube outlet. Thus the temperature of the condensate rain falling down through the tube bundles is not constant but varies along the length of the condenser.

In an actual condenser, this natural value of condensate temperature could represent a significant thermodynamic loss in terms of heat rate. For this reason, some of the vapor leaving the turbine exhaust is arranged to bypass the tube bundles and raise the temperature of the condensate from the natural values calculated using the Nusselt procedure to a value closer to the saturation temperature of the vapor entering the condenser. Without this design feature, the temperature of the condensate will always tend to be lower than that of the exhaust vapor. With this bypass feature, which raises the condensate temperature in the hot well before it is recycled to the boiler, the amount of heat required of the boiler to bring the steam to desired throttle conditions is minimized and the overall heat rate is reduced.

Vapor bypassing also helps to reduce tube bundle pressure drop, thus facilitating the removal of non-condensibles by the air removal system.

1.4 APPARATUS FOR THE LABORATORY STUDY OF HEAT TRANSFER

There has long been a need to quantify the heat transfer properties of metal tubes operating under a variety of conditions, not only to establish sound parameters to be used in the design of an actual condenser, but also to explore the effect on heat transfer of fouling deposits or even of tube coatings or linings. Nusselt [1916] was among the earliest successful investigators, but other apparatus and experimental techniques were used by Wenzel [1962] to establish fundamental clean-tube heat transfer relationships, frequently under the auspices of the Heat Exchange Institute (HEI). The apparatus used in these experiments was quite elaborate and required the use of live steam.

Those involved in the servicing of power-plant condensers are often challenged to quantify the performance improvement that will result from implementing the technology they are offering. One way was to better understand the present state of the tubes in a condenser and the effect of fouling on performance and to determine the best method of bringing the tubes back to a clean condition. Thus, in the early 1990s, it became clear that a more routine method was needed for conducting heat transfer tests on both new and used tubes, but one which also simulated closely the actual condensing conditions under which the tubes operated in practice, including the design water velocity. Several versions of the apparatus illustrated in Figure 1.4 have been built and used successfully to quantify the heat transfer performance of a wide variety of tubes.

The heart of the apparatus is a model condenser in which the tubes to be tested are mounted and up to four tubes can be tested at one time. A small quantity

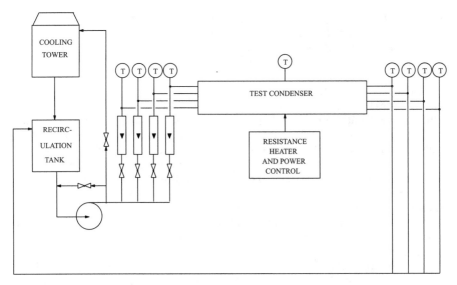

FIGURE 1.4. Heat transfer rack—flow schematic.

of distilled water is placed in the bottom of the condenser; variable-input electric heaters are provided to evaporate the water, while a vacuum pump is used to evacuate any air or non-condensibles present. The cooling water is pumped through a closed system which is also provided with a cooling tower to regulate the temperature of the cooling water supplied to the model condenser. The instrumentation provided includes flowmeters to measure the flow of water to each tube, together with thermistors, used not only to measure the inlet and outlet temperatures of the water passed to each tube but also the vapor temperature inside the shell. All of this instrumentation is connected to a personal computer which is programmed to store the test data and perform the subsequent data reduction.

Since the thermal conditions within the model condenser are not in a state of absolute equilibrium, it is necessary to conduct a test for a period of about 15 minutes in order to obtain consistent results. Even so, the level of uncertainty obtained, using the method outlined in ASME PTC.19.1 [1985], has been found to be 1.05%, based on the accuracy of the instrumentation. Meanwhile, the standard deviation of the U coefficient calculated from each of the 50 or so data sets acquired during a test run was found to be 1.13%. It has also been found that the test results are repeatable. The test results can be used to calculate not only the actual heat transfer coefficient of the tube as measured under test, but also the expected single-tube heat transfer coefficient calculated in accordance with both the HEI and the ASME procedures. Details of the HEI method of estimating condenser performance are found in Chapter 2.

The procedure for testing a tube often consists of several discrete steps. Whether a tube is new or has been removed from a condenser or heat exchanger in the field, the first step is always to establish the heat transfer coefficient of the tube sample(s) as received. The tubes are then subjected to one or more cleaning procedures, and the change of U coefficient is noted after each step. Finally, the tubes are laboratory-cleaned with an acid solution, and the U coefficient is then measured and compared with the expected value as calculated according to the HEI or ASME procedure.

Model condensers have been constructed to accommodate tubes of various lengths. Some are able to accept tube samples as long as 5 ft. However, tubes which are cut out from a condenser are often quite short. For example, tube supports might prevent a long tube sample from being removed, or the wall of the waterbox may constitute an obstruction. For these cases, a shorter model condenser has been constructed, designed to accept tubes which have an effective heat transfer surface equivalent to only 20 inches of tube length. Provided that the temperature sensors are of high accuracy, the consistency of measurement of tube heat transfer coefficients is maintained.

The results of these single-tube heat transfer tests have been put to a wide range of uses, including verification of plant performance testing, assistance in locating the source of a performance loss, optimization of cleaning methods, and performance monitoring of cooling-water chemical treatment methods. Heat transfer testing has also been used to measure the effect on heat transfer of full-length tube liners and coatings [1994]. The single-tube data can also be substituted in condenser performance calculation models [Putman et al. 1996] to determine the effect of the deposits on unit performance.

1.5 HEAT TRANSFER TEST RESULTS ON NEW TUBES

Condenser designers need to have confidence in their chosen method for predicting the heat transfer coefficient of the tubes. Many use the expected value of the single-tube coefficient calculated using the HEI standard [1995], a long-established practice within the power industry but an empirical method. Details and a commentary on this method will be provided later. The alternative ASME [ASME 1983] method is not new. Unfortunately, although based on the summing of thermal resistances, calculated in accordance with Equations (1.1), (1.11), (1.14), and (1.15) and in accordance with modern heat transfer theory, it has not found wide acceptance within the power industry, possibly due to a lack of familiarity. However, its value does take variations in shell-side conditions into account, an important feature for the accurate monitoring of condenser performance. Thus, because of the unfamiliarity of the method, it was important for us to learn how closely the expected value of the heat transfer coefficient calculated by this method corresponded to the effective (measured) value of the heat transfer coefficient calculated from tube operating data in accordance with Equation (1.21).

This comparison was important for several reasons. First, the HEI method takes into account only the operating conditions of inlet water velocity and temperature, whereas the resistance method also responds to variations in the shell-side conditions. Clearly, if the resistance summation method provided a close correspondence with the heat transfer coefficient measured during tests, then using this method to calculate condenser performance would more consistently reflect

Table 1.2. Comparison of Measured and Expected Values of Heat Transfer Coefficients for Clean Tubes

Tube Material	Water Velocity (fps)	Measured U Coefficient	ASME U Coefficient
Titanium 1 in × 22G	3.21	507.64	508.64
	4.59	620.38	624.65
	5.96	701.29	707.98
90-10 Cu-Ni 0.875 in × 18G	4.73	664.67	638.42
	6.77	783.59	763.85
	8.80	884.69	859.94
Admiralty 0.875 in × 18G	4.73	679.52	674.68
	6.77	821.68	818.03
	8.80	930.22	927.35

reality, not only in thermodynamic but also in economic terms. Using the resistance summation method would also allow fouling resistances to be computed with more confidence. Other uses, to be described later, include being able to quantify and distinguish between the losses that can be attributed to tubesheet fouling, and the associated reduction in tube water velocity, as opposed to tube fouling losses caused by deposits alone.

Table 1.2 shows the results of conducting heat transfer tests in the model condenser on tubes of different materials, sizes, and gage and at various tube water velocities. The correspondence between columns 3 and 4 can be considered excellent.

1.6 FINNED TUBES

The main purpose of this section is to provide an interesting example of how the basic resistance summing principle can be applied to one type of tube which did not have a plain outer surface, i.e., had a form of enhanced tube surface. Although not widely used, tubes with enhanced surface area, such as external fins or roped surfaces, have been installed in several power-plant condensers and heat exchangers for several years. Rabas et al. [1991] report on some good experiences with tubes of this type. The improvement in heat transfer is often impressive, although cleaning can present a problem. However, because these tubes are at this time seldom encountered in practice and the prediction of their heat transfer coefficients is less widely understood, monitoring the performance of condensers containing tubes of this type will not be discussed further.

Publication by the Nuclear Regulatory Commission of Generic Letter 89-13 [1989] resulted in nuclear plants' paying much closer attention to the reliability and performance of heat exchangers associated with safety-related equipment, and especially those not in continuous service. This letter became the stimulus to

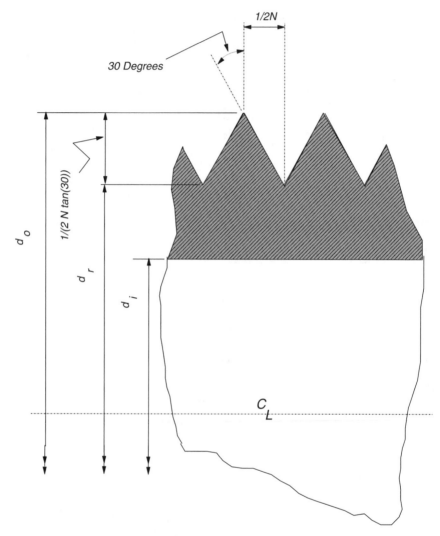

FIGURE 1.5. Finned tube with screw thread surface.

conduct heat transfer tests on tube samples. While plain tubes could be handled in the model condenser described in the previous sections, finned tubes presented a different problem. The following is an example of how the thermal resistance method can be adapted to predict the performance of tubes not commonly found in power plant condensers or heat exchangers.

Consider a tube removed from a water-to-water heat exchanger whose fins were created by cutting a screw thread into the outer surface, as shown in Figure 1.5. Clearly, the original wall thickness of the tube had to be greater than that of a plain tube with the same outside diameter, in order that the fins could be created without weakening the overall mechanical strength of the tube. The adaptation of the thermal resistance method to this application proceeded as follows. For this analysis, let:

d_o = outside diameter of fins
d_r = root diameter
d_e = equivalent outside diameter
d_i = inside diameter
TPI = number of fins per inch
t_e = effective thickness of material

Then

$$d_r = d_o \frac{2}{2\text{TPI} \tan 30°}$$

$$d_e = (d_o + d_r)/2$$

and

$$t_e = (d_e - d_i)/2$$

Since the thermal resistance calculations in the previous sections assume plain tubes, some modifications are necessary when calculating the contributing thermal resistances for finned tubes operating in a condensing environment. The first step is to calculate all thermal resistance with respect to the inner tube surface:

1. Steam-side resistance: Equation (1.1), modified so that the resistance is calculated with reference to the inner surface, becomes:

$$h_f = 0.725 \left(\frac{d_i}{d_o} \right) \left[\frac{12k_f^3 \rho^2 g\lambda}{\mu_f D_e(\Delta T)} \right]^{0.25} \tag{1.23}$$

2. Wall resistance: Equation (1.12) becomes

$$R_w = \frac{d_i}{24k_m} \ln\left(\frac{d_e}{d_i} \right) \tag{1.24}$$

3. Water-side resistance: Equation (1.14) must revert to its original form, in which the film resistance is calculated with reference to the inner surface; thus:

$$R_t = 0.045057 \left(\frac{\mu^{0.373}}{\rho^{0.835} k^{0.538} C_p^{0.462}} \right) \left(\frac{d_i^{0.165}}{V^{0.835}} \right) \tag{1.25}$$

The sum of these resistances is then computed and its reciprocal taken. This is the value of the U coefficient (U_i) calculated with respect to the inner surface. To calculate the value U_{corr} with respect to the outer surface U_o:

$$U_o = U_i \frac{d_e}{d_i}$$

To demonstrate the accuracy of the method, an acid-cleaned finned Cu-Ni tube, configured in accordance with Figure 1.5, was found to have a U coefficient of 738.36 Btu/(ft2 · h · °F), whereas the ASME value which was calculated using the foregoing procedure was 742.73.

1.7 EFFECT OF NON-CONDENSIBLES

Condensers in fossil-fired and nuclear plants are provided with air-removal equipment which allows the concentration of air in the vapor space surrounding the tubes to be kept to an allowable minimum. Condensers are designed to allow for a limited amount of non-condensibles to continuously enter the vapor space. However, if an excessive amount of air leaks into this space or is allowed to accumulate within it, the condenser tubes can exhibit a severe reduction in heat transfer capacity, accompanied by a rise in back pressure. Thus, in fossil and nuclear plants, a great effort is made to avoid the buildup of high concentrations of non-condensible gases.

However, steam generated geothermally often contains large amounts of non-condensible gaseous impurities such as carbon dioxide and hydrogen sulfide which pass to the condenser with the exhaust steam and increase the thermal resistance of the vapor/condensate film on the shell side of the tubes. Quantitative data on the increased thermal resistance caused by such gases is difficult to find in the literature; the model condenser outlined in this chapter was used to study the problem for the geothermal case.

Argon is a gas having a molecular weight close to both carbon dioxide and hydrogen sulfide. Because it is readily available, it was chosen as the gas to be used for these tests. After mounting a cleaned tube in the model condenser and running with water vapor only in the condensing space, it was possible to establish the reference U coefficient. Subsequently, controlled and known quantities of argon were injected into the condensing space and the new value of the U coefficient was found for each incremental concentration of argon. The results are plotted in Figure 1.6.

The experiment confirmed that the presence of a non-condensible gas tends to reduce the tube heat transfer coefficient but also showed that quite small concentrations can cause a serious reduction in heat transfer rate. The condensers

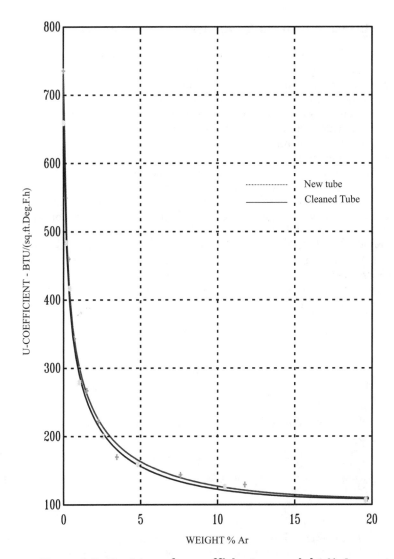

FIGURE 1.6. Heat transfer coefficient vs. weight % Ar.

installed in geothermal plants are provided with a larger surface area in order to account for the higher film-side thermal resistance. However, in such plants, fouling on the inside surfaces of the tubes can still tend to increase back pressure and reduce the amount of power which could be potentially generated from a given quantity of geothermal steam. Thus, even though the effect of the non-condensible gas has been largely offset by providing a more generous tube surface area than would otherwise have been the case, the economic impact of fouling can still be serious and should continue to be monitored.

1.7.1 Effects of Air on Condensate Film Heat Transfer

With no non-condensible gases present, the driving force for heat transfer is $(T_v - T_b)$. However, when they are present, Sparrow et al. [1967] suggest that the reason for their causing a reduction in shell-side film heat transfer coefficient is that their presence lowers the corresponding partial pressure of the vapor. This, in turn, lowers the interface saturation temperature at the surface of the condensate (or saturation temperature corresponding to the partial pressure of the vapor, T_{pp}) so that the driving force, now $(T_{pp} - T_b)$, is also lowered and thereby decreases the heat transfer rate.

Quantitative information in the literature is not easy to apply to practical condenser situations. Silver [1963–1964] postulated a thermal resistance for the interface between the outer surface of the tubes and the general volume of exhaust vapor. However, Sparrow et al [1967] threw doubt on the concept of interface resistance, and the literature has been silent on the matter since then.

Henderson and Marchello [1969] also studied the problem and provided correlations between the mole percent of air in the vapor (Y) and the film coefficient ratio H, this being the effective film heat transfer coefficient H_M at the time of the test divided by the original Nusselt film coefficient H_{Nu} calculated for the same conditions (see Equation (1.1)), so that:

$$H = \frac{H_M}{H_{\text{Nu}}} \qquad (1.26)$$

The correlation they found between parameters H and Y was:

$$H = \frac{1}{1 + 0.511Y} \qquad (1.27)$$

Their paper includes curves not only for air/water vapors but also toluene and nitrogen. Figure 1.7 reproduces their plot of H and Y for air. Even though Figure 1.6 is not plotted on the same basis, there is still a strong similarity between it and Figure 1.7.

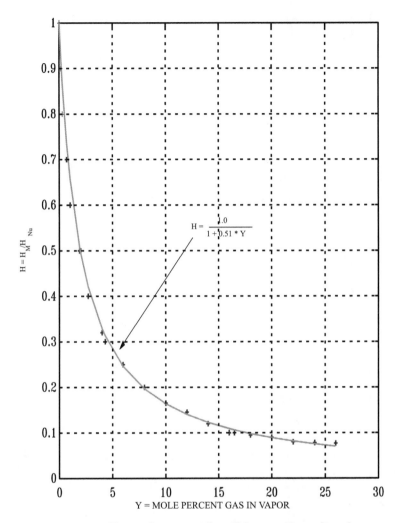

FIGURE 1.7. Effect of non-condensibles on Nusselt value.

Chapter 2

CONDENSER PERFORMANCE MONITORING

2.1 PAST PRACTICE

Programs to monitor equipment performance in both nuclear and fossil power plants were written for even some of the early process computers used in the electric utility industry, but these first-generation machines were difficult to program and slow in execution, and the performance calculation procedures therefore often had to be simplified. When on-line computer monitoring was mandated after the Three Mile Island accident in 1979, the programs became more sophisticated, but microprocessors were still slow. Steps had to be taken to reduce execution time. For instance, the algorithms to calculate the properties of steam and water could not use the original FORTRAN code published by ASME [1993] in the contemporary editions of "The Steam Tables"; the data had to be generated from this code and then regressed. Most of the resulting algorithms consisted of families of third-order piecewise polynomials; the appropriate polynomial being selected according to the temperature or pressure range. Because of these practical limitations, the scope of the performance monitoring programs was still narrow.

The introduction of distributed control systems in 1982 allowed distributed processing of substantially increased volumes of data, which could now be scanned in real time and without affecting the system duty cycle. Certain microprocessor-based devices could be dedicated to performance monitoring, and as a result, the quality and complexity of the programs greatly improved. Within these systems, the condenser performance monitoring task normally consisted of two main features:

1. Calculation of the HEI cleanliness factor, discussed later in this chapter
2. Calculation of heat-rate deviation as a function of deviation of back pressure from its design value

The HEI method allows calculation by hand and works well with condensers having a single compartment. However, for a condenser designed with multiple compartments, each operating at a different back pressure, there was usually insufficient instrumentation, or the design prevented access to intercompartmental water temperature values. Thus, the results with multicompartment condensers were often questionable. Furthermore, recent studies have found that the design HEI cleanliness factor may not be a constant but seems to vary with load, falling in value as the load drops. Thus, under cycling operation, plots of the HEI cleanliness factor may not provide a clear picture of the degree to which the condenser may have become fouled. For these and other reasons, it is difficult to convert HEI cleanliness factors to a reliable cost equivalent of fouling losses.

Heat-rate correction curves for deviations in back pressure from the design value are normally a part of the set of thermal kit data provided by the turbine vendor at the time of equipment acceptance. An alternative form of these curves show the percent correction which should be applied to *generated power* as a function of the back-pressure deviation.

A typical heat-rate correction curve is shown in Figure 2.1, in which the relationship between heat-rate correction and back pressure is plotted for various values of turbine throttle flow. Figure 2.2 shows another way of presenting the information: the heat-rate correction factors are plotted against condenser flow for various values of condenser back pressure. Figure 2.3, on the other hand, shows a plot of the load correction associated with condenser flow, for various values of condenser back pressure. From Figures 2.2 and 2.3 it can be seen that dropping the back pressure from 2.5 to 0.5 in-Hg can cause the heat rate to drop by almost 9% or allow the load to rise by approximately 10%.

One of the features of modern performance monitoring is the controllable losses display, whose purpose is to indicate the economic impact of deviations from design value in several measures of equipment performance—among them deviations in boiler efficiency from that expected for the load, as an hourly cost; heat-rate corrections as a function of throttle temperature and pressure deviations; heat-rate corrections as a function of reheater temperature and pressure drop deviations; and heat-rate correction in response to back-pressure deviations. For the first four items, the question is, How much can these costs be reduced by improved control of the boiler? However, in the case of back-pressure deviations, it is not so clear where to look for improvement. Even in a clean condenser, some of the deviation is a natural response to changes from design in the cooling-water inlet temperature and/or flow. Only some of the deviation may be due to fouling and/or air ingress, the consequences of which can normally be improved by performing some maintenance activity. But exactly how much of a given back-pressure change can be attributed to fouling and/or air ingress is an important factor in the economics of condenser maintenance decisions.

2.2 PERFORMANCE MONITORING AS AN EXTENSION OF EQUIPMENT ACCEPTANCE TESTS

Two organizations—the Heat Exchange Institute (HEI) and the American Society of Mechanical Engineers (ASME)—have published standards in the United States for use not only as the basis for condenser design but also for equipment acceptance tests. The computational methods outlined in these standards have also been used as the basis for periodic or continuous monitoring of condenser performance. The HEI has published several versions of its *Standard for Steam Surface Condensers* over the past years, the latest being dated 1995 [HEI 1995]. The details for calculating the expected *tube-bundle* heat transfer coeffi-

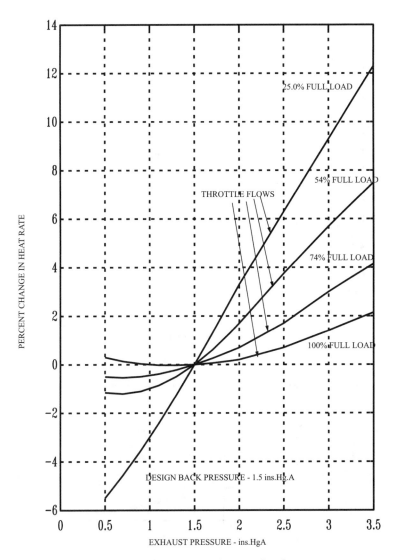

FIGURE 2.1. Heat-rate correction vs. back pressure.

cient change from one edition to the next, and care must be taken when comparing the cleanliness-factor values calculated using, say, the 1982 standard with the results using the 1995 standard for the same set of test data.

The American Society of Mechanical Engineers has also developed standards. Its PTC.12.2-1983 [ASME 1983] recommended the use of the thermal resistance method to calculate the expected *single-tube* heat transfer coefficient. Unfortunately, this standard has rarely been used as the basis for a condenser acceptance

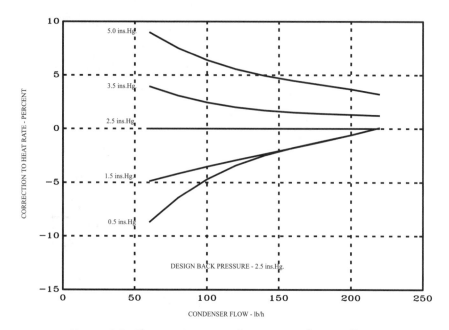

FIGURE 2.2. Heat-rate correction vs. condenser flow.

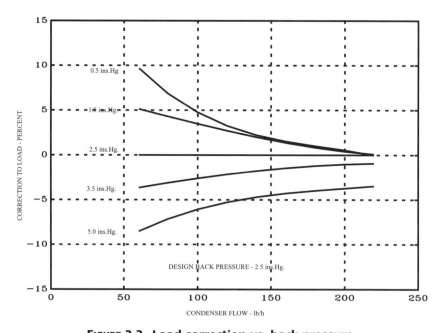

FIGURE 2.3. Load correction vs. back pressure.

test procedure, probably because the thermal resistance calculations required a computer program for their evaluation. This version of the standard was also largely silent as to exactly how it should be used to monitor the performance of condensers operating under partial-load conditions.

Tsou [1994] also explored the use of the ASME thermal resistance method for calculating the expected single-tube heat transfer coefficient as a function of tube operating conditions and physical dimensions, but his approach does not appear in the final version of the current standard, PTC.12.2-1998 [ASME 1998]. In the new standard, the methods for calculating the tube-wall and water-film thermal resistances have been updated, and these are contained in Equations (1.11) and (1.14) of Chapter 1. However, this standard is, once more, designed as an equipment acceptance test procedure rather than a guide to continuous performance monitoring.

2.3 THE HEI METHOD FOR MONITORING CONDENSER PERFORMANCE

The series of *Standards for Steam Surface Condensers*, published over the years by the HEI, all calculate the reference or expected value of the *overall tube-bundle heat transfer coefficient*, as a function of the following:

- Base U coefficient as a function of tube outside diameter and water velocity (UBASE)
- Cooling-water inlet temperature (F_W)
- Tube material and gage factor (F_M)
- Cleanliness factor (F_C)

In the earlier standards, UBASE and F_W were obtained from curves contained in the standards, but the 1995 version also contains tables of values which can be either interpolated or regressed. In all standards, the material factor F_M was provided in the form of a table. The following relationship is used to calculate the expected overall tube-bundle heat transfer coefficient:

$$U_{\text{ref}} = \text{UBASE} \times F_M F_W F_C \qquad (2.1)$$

For continuous monitoring of condenser performance, it is clearly possible to write computer programs which perform table lookup functions, interpolating the data between adjacent points in the table. However, continuous monitoring can be better performed using continuous functions and, indeed, the very act of regressing data can reveal insights into the thought processes behind the data arrangements. The 1995 HEI standard was drawn up with this in mind.

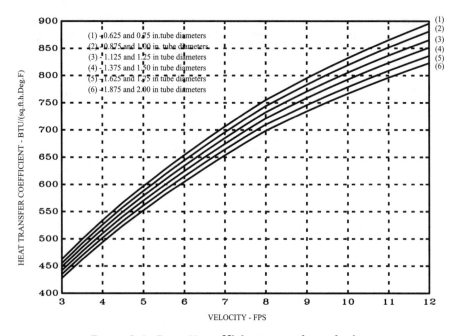

FIGURE **2.4. Base *U* coefficient vs. tube velocity.**

2.3.1 Base *U* Coefficient

A table in the new HEI standards contains the values of the base or uncorrected heat transfer coefficient for velocities between 3 and 12 ft/s and six tube sizes ranging from 0.625 through 2 inches outside diameter. This data is plotted in Figure 2.4. The eighth edition of these standards [HEI 1984, 1989] covered only the velocity range 3 through 8 ft/s and was fairly easy to regress with a close fit. However, the same degree of fit could not be obtained with the increased range covered in the new standards. In the eighth edition, each of the six tube sizes had its own constant assigned to it for use in the equation

$$U = C\sqrt{V} \tag{2.2}$$

and it was also found that the relationship between this constant and tube diameter followed a linear law, such that:

$$U = (278 - 16d_o)\sqrt{V} \tag{2.3}$$

Figure 2.5 is a plot of the constants calculated from the table contained in the 1995 standard for various tube sizes. Between 3 and 8 ft/s, the constant was maintained at a steady value, but between 8 and 12 ft/s, it declines linearly as the

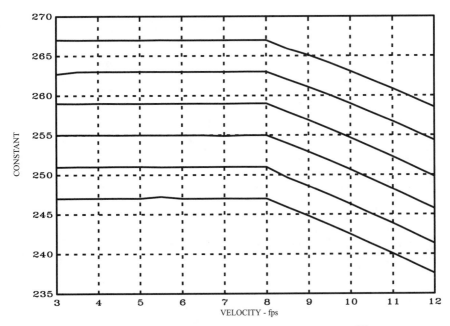

FIGURE 2.5. Constant in Equation (2.2), $U = C\sqrt{V}$.

velocity increases. The regression analysis of this data showed that two equations were needed to calculate the base U coefficient, one for tube velocities below 8 ft/s and the other for velocities greater than this threshold; these equations are defined as follows:

```
IF (VELOCITY .LE. 8 fps) THEN
    U = (278 - 16 * d₀) * SQRT (V)
ELSE                                                                    (2.4)
    U = (2.78.81395 - 16.532697 * d₀ - 2.273194
        * (V - 8.0)) * SQRT (V)
END IF                                                                  (2.5)
```

We looked for an explanation for the discontinuity shown in Figure 2.5, by comparing the HEI data with equivalent data generated using the Rabas/Cane correlation [Rabas 1983] of Equation (1.14). Assuming that the tube wall resistance R_w and steam-side thermal resistance R_s are givens, then, using the Rabas-Cane correlation, the heat transfer coefficient as a function of velocity can be calculated from

$$U = \frac{1.0}{f\left(V^{-0.835}\right) + R_w + R_s} \tag{2.6}$$

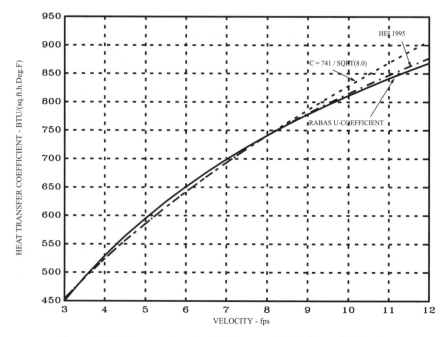

FIGURE 2.6. Comparison of Rabas and HEI *U* coefficients.

Clearly this does not have the same form as the HEI equation, (2.2). Consider a 1-in-diameter 18G admiralty tube provided with cooling water having an inlet temperature of 70°F and a velocity of 8 ft/s as the reference. A vapor temperature of 89°F is also assumed. The wall and steam-side thermal resistances can be calculated using this information.

Figure 2.6 shows the correlation between the U coefficient calculated using the Rabas-Cane correlation as embodied in equation (2.6) and the HEI values if, following the 1989 version of the standard, the 1-in tube constant of $C = 262 = 741/\sqrt{8.0}$ is extended into the region 8 through 12 ft/s. Although Figure 2.6 shows some deviation over the range 3–8 ft/s, a much larger discrepancy is apparent over the range 8–12 ft/s. However, if the constant is modified in accordance with Equation (2.5), reflecting the declining value associated with the velocity range 8–12 ft/s, a much closer correlation is obtained. This suggests that the discontinuity introduced by the HEI in 1995 and shown in Figure 2.5 results from allowing the values of the HEI heat transfer coefficients to reflect more closely the known effect of velocities above 8 ft/s on the water-side thermal resistance, as embodied in the Rabas-Cane correlation.

2.3.2 Temperature Correction Factor

Regression analysis of the temperature correction factor data contained in the HEI standard indicates that the best fit is obtained by breaking the water temper-

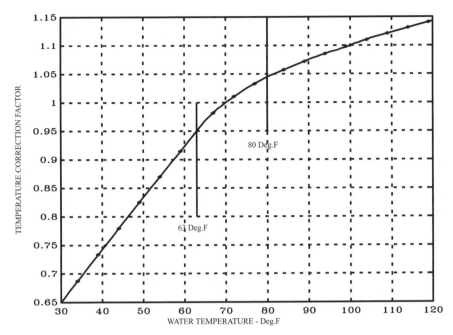

FIGURE 2.7. HEI temperature correction factor, based on 9th edition of the HEI standard [1995].

ature range of 30 through 120°F into three segments, each having its own third-order piecewise polynomial. Expressed in FORTRAN code, if T is the cooling-water inlet temperature for which the temperature correction factor (TCF) is to be found, then:

```
IF(T .LE. 63.5) THEN
    TCF = 0.36536 + 9.552E-03 * T - 1.8625E-07 * T**2
        - 6.689E-08 * T**3
ELSE IF(T .LE. 80.0 .AND. T .GT. 63.5) THEN
    TCF = -2.7374 + 0.13694 * T - 1.7032E-03 * T**2
        + 7.2813E-06 * T**3
ELSE IF(T .GT. 80.0) THEN
    TCF = 0.63558 + 7.4478E-03 * T - 3.3738E-05 * T**2
        + 5.7428E-08 * T**3
END IF
```

A plot of these polynomials against the original data is shown in Figure 2.7

Of interest is how this temperature correction factor might correspond with that obtained using the Rabas-Cane correlation. We fixed the water velocity at

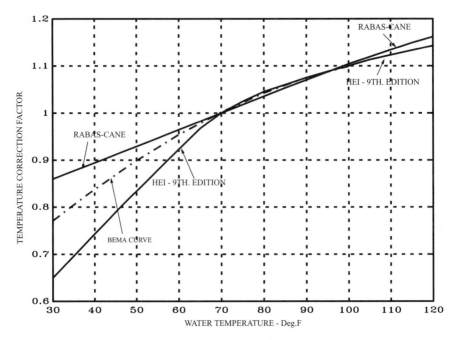

FIGURE 2.8. Tempterature correction factor vs. temperature.

7.0 ft/s, used the values of R_w and R_s as before, and allowed the inlet water temperature to vary. The temperature correction factor at 70.0°F is assumed to be unity. Figure 2.8 compares the HEI values of the temperature correction factor with those obtained using the Rabas-Cane correlation. While there is quite good correspondence over the temperature range 70.0–120°F, the correlation below 70.0°F leaves much to be desired. Clearly, the values given by the HEI will tend to reduce the expected heat transfer coefficient at lower water inlet temperatures. It is of interest that Taborek [1992] compared a plot of the HEI values with those included in the 1967 BEMA (British Electrical Manufacturers Association) standard, but without offering any explanation for what seems to be a similar discrepancy.

2.3.3 Material Correction Factor

Table 2.1 excerpts some of the HEI material correction factors assigned to various tube materials and wall thicknesses. For a given material, the relationship between material correction factor and wall thickness is not constant, but note that these are factors applied to an overall tube-bundle heat transfer coefficient.

Table 2.1.   HEI Material Correction Factors FM

	BWG (in)						
Material	24 (0.022)	22 (0.028)	20 (0.035)	18 (0.049)	16 (0.065)	14 (0.083)	12 (0.109)
Admiralty metal	1.03	1.02	1.01	1.00	0.98	0.96	0.93
Arsenical copper	1.04	1.03	1.03	1.02	1.01	1.00	0.98
Copper iron 194	1.04	1.04	1.03	1.03	1.02	1.01	1.00
Aluminum brass	1.02	1.02	1.01	0.99	0.97	0.95	0.92
Aluminum bronze	1.02	1.01	1.00	0.98	0.96	0.93	0.89
90-10 Cu-Ni	0.99	0.98	0.96	0.93	0.89	0.85	0.80
70-30 Cu-Ni	0.97	0.95	0.92	0.88	0.83	0.78	0.71
Cold Rolled carbon steel	1.00	0.98	0.97	0.93	0.89	0.85	0.80
SS type 304/316	0.90	0.86	0.82	0.75	0.69	0.62	0.54
Titanium	0.94	0.91	0.88	0.82	0.77	0.71	0.63

2.3.4   Design Cleanliness Factor

In the design of a condenser for a particular installation, a data sheet is prepared which gives the most difficult set of operating conditions under which the condenser is expected to function and the amount of latent heat it is required to remove from the turbine exhaust. The operating conditions include the cooling-water inlet temperature, cooling-water flow, and, usually, cooling-water outlet temperature as well. From this set of design operating conditions, the corresponding effective overall heat transfer coefficient of the condenser, U_{eff}, can be calculated, using a transformation of Equation (1.22):

$$U_{eff} = \frac{Q}{A \times LMTD} \qquad (2.7)$$

The reference overall tube-bundle heat transfer coefficient U_{ref} calculated in Equation (2.1) can be considered to be composed of two parts:

- The value expected as a function of the physical properties of the tube, water velocity, and water inlet temperature
- A multiplier F_c which, for a clean condenser, is termed the *design cleanliness factor*

An analysis of design data sheets prepared by condenser manufacturers shows that a design cleanliness factor is selected such that

$$U_{eff} = U_{ref}$$

or

$$F_c = \frac{U_{\text{eff}}}{\text{UBASE} \times F_M F_W} = \frac{U_{\text{eff}}}{U_{\text{HEI}}} \tag{2.8}$$

One reason for the discrepancy between U_{eff} and U_{HEI}, i.e. prior to the application of the design cleanliness factor, can be attributed to the small amount of air always present within the condenser shell no matter how effective the air-removal system. But the main reason is the distribution of heat transfer conditions between the uppermost tube bundles, which receive the initial brunt of high exhaust velocity and vapor dryness, and lower bundles on which the condensate rain falls and which are exposed to vapor with generally higher moisture contents and lower velocities. Other factors which affect the value of U_{ref} prior to the application of F_c include:

- Condensate loading effects
- Non-condensible gas buildup
- Saturation temperature depression due to vapor-side pressure drop
- Vapor velocity effects

A design cleanliness factor may be thought of as a multiplier the condenser designer uses to discount the theoretical value of the overall tube-bundle heat transfer coefficient, i.e., the product UBASE *times* $F_M F_W = U_{\text{HEI}}$, in order that the duty on the condenser can be satisfied under the desired set of operating conditions. It is, of course, up to the designer to ensure that these conditions are fulfilled by providing tubes of the appropriate length (i.e., surface area) as well as adequate vapor lanes through and around the tube bundles so that all the low-pressure exhaust vapor is condensed at the desired back pressure.

Design cleanliness factors of 85% are very common, although, with some condensers equipped with stainless steel tubes, design cleanliness factors of 90% or higher can also be encountered. Meanwhile, condensers associated with geothermal plants often have design cleanliness factors between 65% and 70% in order to offset the effect of the large concentration of non-condensible gases which accompany steam obtained from geothermal sources.

It cannot be emphasized enough that *design* cleanliness factors contain no attributes related to fouling, because they apply only in the context of a *clean* condenser, and one subjected only to a stated set of operating conditions, which include cooling-water inlet temperature and flow. However, cleanliness factors calculated after the unit has been in operation will be somewhat lower than the design cleanliness factor since some fouling deposition may now occur. Unfortunately, HEI overall tube-bundle U coefficients are not able to reflect changes in the shell-side conditions, so that it is difficult to directly correlate changes in HEI cleanliness factors with fouling resistances. Even so, monitoring the values of cleanliness factors calculated in accordance with the HEI method can provide useful information about the changes occurring within the condenser and is a practice which has found wide industry acceptance.

2.3.5 Design Cleanliness Factor and Nusselt Theory

Some further support for the need for a design cleanliness factor can be gained from the original paper by Nusselt [1916], who considered condensation as being filmwise in nature, cascading from row to row down a bank of tubes through viscous action under gravity, and free from turbulence. In reviewing Nusselt's work, Short and Brown [1951] and, later, Kern [1958] observed that the condensate film heat transfer coefficient for the pth tube in a bank of n tubes can be found from the following relationship, where h_1 is the Nusselt single-tube heat transfer coefficient, which applies only to the first row:

$$h_p = h_1[p^{3/4} - (p-1)^{3/4}] \qquad (2.9)$$

which, after summing h_p over the range $p = 1$ to n, gives:

$$h_n = h_1 \frac{n^{3/4}}{n} = h_1 n^{-0.25} \qquad (2.10)$$

It can be shown that the term $p^{3/4}$ in Equation (2.9) is the cumulative value of the h_1 multiplier summed from row 1 thru row p, while $(p-1)^{3/4}$ is the cumulative value summed from row 1 through row $(p-1)$. Thus, the difference between the two terms in Equation (2.9) is the value of the h_1 multiplier at row p.

The author has calculated values generated by Equation (2.9) and has compared them with equivalent values calculated using Equation (65b) in Nusselt's original paper [1916], the mean difference between them being 0.77% over 100 rows of tubes. Thus, Equation (2.9) provides effective values of row film heat transfer coefficients without having to evaluate the original and complicated Nusselt equation.

From Equations (2.9) and (2.10) it will be seen that, when $p = 1$, the value of h_p is the Nusselt single-tube value h_1; but when $p = 100$, h_p has a value of $0.24h_1$, indicating a reduction in the Nusselt single-tube value from row to row down the height of the tube bank. The overall effect with 100 rows is $h_n = 0.32h_1$.

Nusselt considered the condensate to be in viscous flow, saying that it "descends from one tube to the tube below as a continuous sheet without disturbing the condensate on the tube below." Kern insisted that the condensate drains not as a sheet but as droplets or, at high loadings, as continuous stream points, either effect tending to disrupt the purely viscous film concept. Thus, for the case when turbulence is present, Kern [1958] examined the data published by others and proposed an adjustment to Equation (2.10) to reflect deviations from purely viscous flow:

$$h_n = h_1 n^{-0.16667} \qquad (2.11)$$

To investigate the consequences of Equations (2.10) and (2.11), it is useful to base the analysis on heat transfer coefficients calculated using the thermal resistance method. A quantity known as the *performance factor* was therefore introduced by Tsou [1994]. Assuming that the value of U calculated in Equation (1.15) is designated U_{ASME}, then performance factor PF may be defined as:

$$PF = \frac{U_{eff}}{U_{ASME}} \qquad (2.12)$$

i.e., the single-tube equivalent of cleanliness factor F_c related to U_{HEI} in Equation (2.8).

We now take the case of a condenser with 100 rows of tubes, having an overall single-tube heat transfer coefficient of 800 Btu/(ft2 · h · °F), and assume that the condensate-film thermal resistance is 27% of the total resistance, a typical situation. Equation (2.11) gives only the film heat transfer coefficient, so that, to estimate the design performance factor corresponding to Equation (2.11), the following calculations must be performed:

Total tube thermal resistance
$= 1/800 = 1.25 \times 10^{-3}$ °F/[Btu/(ft^2 · h)]
Film thermal resistance
$= 0.27 \times 1.25 \times 10^{-3}$
$= 0.3375 \times 10^{-3}$ °F/[Btu/(ft^2 · h)]
$= 2963.0$ Btu/(ft^2 · h · °F)
Adjusted film U coefficient
$= 2963 \times 100^{-0.25}$
$= 2963 \times 0.31623$
$= 936.98$ Btu/(ft^2 · h · °F)
$= 1.0673 \times 10^{-3}$ °F/[Btu/(ft^2 · h)]
Residual tube thermal resistance
$= (1 - 0.27) \times 1.25 \times 10^{-3}$
$= 0.9125 \times 10^{-3}$
Adjusted film U coefficient
$= 2963 \times 100^{-0.16667}$
$= 2963 \times 0.46416$
$= 1375.3$ Btu/(ft^2 · h · °F)
$= 0.72711 \times 10^{-3}$ °F/[Btu/(ft^2 · h)]
Residual tube thermal resistance
$= (1 - 0.27) \times 1.25 \times 10^{-3}$
$= 0.9125 \times 10^{-3}$
Adjusted tube thermal resistance
$= 0.9125 \times 10^{-3} + 0.72711 \times 10^{-3}$
$= 1.6396 \times 10^{-3}$ °F/[Btu/(ft^2 · h)]
Adjusted tube U coefficient
$= 609.9$ Btu/(ft^2 · h · °F)
Single-tube design performance factor
$= 100 \times 609.9/800$
$= \underline{76.238\%}$

Thus, the performance factor derived using the Kern adjustment is 76.238%, which compares with the design performance factor typically obtained with ASME single-tube coefficients of 74%, equivalent to an HEI design cleanliness factor of 85%. This procedure shows how the original Nusselt theory for calculating the single-tube coefficient for tubes arranged in rows, Equation (2.10), can be used for the case where the flow is nonturbulent, i.e., equivalent to low exhaust flow conditions; while the Kern adjustment of Equation (2.11) represents the behavior with turbulent flow under design (i.e., full-load) conditions. Together they provide a rather simple but plausible explanation for the need to apply a fundamental "cleanliness factor" to the design of a condenser tube bundle, while the difference in the exponents in these two equations suggests one reason why the design cleanliness factor can change with load.

2.4 RELATIONSHIP BETWEEN CLEANLINESS FACTOR AND LOAD

Traditionally, the design cleanliness factor has been thought of as a constant which applies at all loads while the condenser remains clean. It has also been assumed that cleanliness factors are less than the design value only when the condenser has become fouled. However, recent research by Putman and Karg [1999] suggests that this may not be the case.

Consider a steam turbogenerator unit with a condenser which has just been cleaned, the load on the unit subsequently being varied over its full operating range. During this exercise, the cooling-water flow remained unchanged while the inlet temperature also remained sensibly constant. Figure 2.9 is a typical plot of the results obtained from this exercise; it shows that the cleanliness factor varies almost as a linear function of load, even for a clean condenser. This finding is important to cycling plants and can affect the way in which apparently low cleanliness factors under partial load conditions are judged; in other words, under cycling conditions, a low cleanliness factor is not necessarily an indication that fouling has occurred.

Let us consider why the cleanliness factor should change even though the cleanliness of the condenser has not changed. A condenser cleanliness factor (CF) is defined as:

$$CF = 100 \frac{U_{eff}}{U_{HEI}} \tag{2.13}$$

Because the water flow and temperature were sensibly constant during the test period, the value of U_{HEI} would be essentially constant. Thus, for CF to change, it is necessary for the effective heat transfer coefficient U_{eff} to change. From Equation 2.7, as the duty falls, the associated drop in cooling-water outlet temperature will cause the LMTD to rise slightly, but these changes do not fully explain the behavior. One possible explanation is suggested in Section 2.3.5. For a condenser with 100 rows and a single-tube U coefficient of 800 Btu/(ft^2 · h · °F), using Equation (2.10) gives a performance factor of 63% with nonturbulent flow,

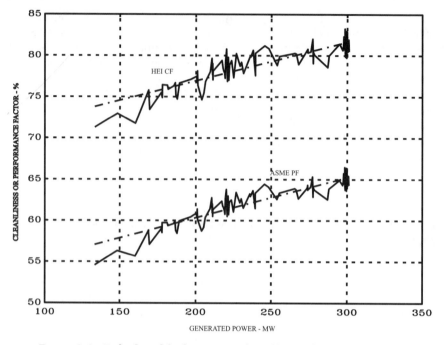

FIGURE 2.9. Relationship between cleanliness factor and load.

while a performance factor of 76.23% with turbulent flow is obtained if Equation (2.11) is used. Since Equation (2.10) more closely approaches part-load conditions while Equation (2.11) should be used when approaching full load, the difference between the exponents in these two equations suggests one reason why the design performance factor and, hence, design cleanliness factor can change with load.

2.5 THE THERMAL RESISTANCE METHOD FOR MONITORING CONDENSER PERFORMANCE

The limited amount of information contained in HEI cleanliness factors has long been recognized. In Appendix 2 of the 1983 edition of the ASME Power Test Code for Steam Surface Condensers [ASME 1983] there was an attempt to introduce the concept of an expected single-tube heat transfer coefficient based upon tube-wall, water-, and steam-side film thermal resistances calculated as a function of design and/or current operating conditions. However, the forms of these calculations proved to be more complicated than the industry was prepared to accept for routine use, and the technique languished. In our own investigations involving the routine heat transfer testing of both new tubes and tubes removed for detailed examination, we have successfully used the 1983 ASME thermal resistance method to verify the measured heat transfer coefficient, and it has become a central feature of the condenser performance models to be described. Support for the use of the thermal resistance method was provided in 1992 by Taborek and Tsou [1992], and Tsou [1994] outlined a broader scheme for applying the technique for condenser performance monitoring.

Just as a design cleanliness factor is applied within the HEI method to discount the distributed heat transfer effects across tube bundles, Tsou's "performance factor" must be applied to expected single-tube heat transfer coefficients calculated from thermal resistances. These performance factors are found to be smaller than the HEI design cleanliness factors; for example, if the HEI design cleanliness factor is 85%, the equivalent performance factors would lie between 70 and 73%. The generally lower values for performance factors can be attributed to the HEI U coefficients' being overall tube-bundle values whereas the U coefficients calculated from thermal resistances are single-tube values.

Reference was made earlier to the recently published PTC.12.2-1998 [ASME 1998]. Unfortunately, this seems to have been written primarily as a condenser acceptance test procedure and contains few references to or insights into the procedures to be followed for on-line performance monitoring. The new code does, however, contain the improved method for calculating tube wall resistance (Equation (1.11)) as well as the Rabas/Cane correlation for water film resistance (Equation (1.14)).

2.5.1 Applying a Performance Factor to the Single-Tube Heat Transfer Coefficient

The term *performance factor* (PF) was defined in Equation (2.12). It is the multiplier applied to the single-tube heat transfer coefficient, calculated according to Equation (1.15), so that it can be reconciled with the effective heat transfer coefficient U_{eff}, calculated for the same set of operating conditions and using the Fourier equation (1.21). As mentioned, because U_{HEI} is an overall tube-bundle U coefficient, the PF is numerically smaller than F_c. In both cases the multiplier is applied directly to the corresponding reference value; the performance models described in later chapters use the PF in this manner. However, when a *design* performance factor is used as a multiplier, it applies an equal weight to all three thermal resistances, i.e., shell-side and water-side film resistances and tube-wall resistance. Yet the values of the last two are identical in both their effective and expected U coefficients. Thus, the use of an overall multiplier to effect the data reconciliation may not be the most appropriate approach.

Another way of making the necessary reconciliation, and in accordance with modern heat transfer theory, is to apply the multiplier only to the Nusselt term, h_f, of the single-tube heat transfer equation (1.15) for a clean tube (after setting R_{foul} to zero); thus,

$$U_{ASME} = \frac{1}{\dfrac{1}{PF_h \times h_f} + R_w + R_t} \tag{2.14}$$

One justification for this approach is contained in Equations (2.10) and (2.11), in which the condensate film U coefficient is calculated by modifying the Nusselt value for the first row by a multiplier which is a function of the total number of rows in the tube bundle. Further, in 1991 Taborek and Tsou [1992]—after outlining the history of condenser performance monitoring techniques, starting with Orrok's method and then discussing the HEI method based on Wenzel's data—proposed a new prediction method using thermal resistances. In the course of their discussion they recognized the distribution of the steam-side heat transfer effect throughout the tube bundles, lowering its value to what they termed $(h_c)_{eff}$, and proposed a modification to the basic Nusselt equation for steam-side film heat transfer $(h_c)_{Nu}$ in the form of a correction factor:

$$PF_h = \frac{(h_c)_{eff}}{(h_c)_{Nu}} \qquad \text{(Taborek's Equation 15)}$$

Developing this line of thought further, Tsou [1994] proposed to calculate the value of the film-side thermal resistance by subtracting the tube-wall and water-side thermal resistances from the reciprocal of the overall heat transfer coefficient, $1.0/(h_c)_{eff}$:

$$R_s = \frac{1.0}{U_{eff}} - R_m - R_t \left(\frac{d_o}{d_i} \right) \qquad (2.15)$$

He illustrated the method in the equation included in Section 8 of his Appendix. This is very similar to the method adopted in Equation (5.1.12) of the current PTC.12.2-1998, although in the latter the measured fouling resistance is also subtracted. The importance of the Taborek/Tsou insight is that, while the traditional approach has been to consider the design and effective heat transfer coefficients as being related by a multiplier (cleanliness or performance factor), in actual fact it is only the steam-side film heat transfer coefficient which should be related to a (different) multiplier.

Note that the value of PF_h can be calculated from the condenser design data, after first using Equation (2.15) to calculate the value of R_s for both the reference single-tube and the overall heat transfer coefficients. It will require further study to determine how applying the multiplier to the condensate-film heat transfer coefficient, rather than to the overall single-tube U coefficient, will affect the models and how a multiplier used in this way will change as the load changes. If the overall effect on condenser losses is small, then introducing this added complication may be unnecessary. But purists will probably insist that models should reflect the true nature of the heat transfer mechanisms involved.

Table 2.2. Condenser Design Data

Measure	Admiralty	Admiralty	70-30 Cu-Ni	AL-6X	Units
			Material		
Tube outside diameter	1.0	1.0	0.875	0.875	in
Tube wall	0.049	0.049	0.049	0.028	in
Length	26.25	30.0	28.25	42.25	ft
No. of tubes	4040	7056	7790	17,824	
No. of passes	2	2	1	1	
Thermal conductivity	70	70	18	7.9	Btu/(h · ft · °F
Material correction factor	1.0	1.0	0.88	0.85	
Cooling-water temp. in	75.0	85.0	70.0	80.0	°F
Cooling-water temp. out	89.83	101.5	80.2	91.7	°F
Back pressure	1.74	2.415	1.5	2.0	in-Hg
Cooling-water flow	29,000	55,000	85,000	249,200	GPM
Exhaust flow	229.67	477.43	402.86	2062.44	klb$_m$/h

2.6 CALCULATIONS BASED ON CONDENSER DESIGN DATA SHEETS

Tables 2.2 and 2.3 summarize the calculations which are commonly performed in creating and/or evaluating the internal consistency of condenser design data sheets. Experience has shown that the information is not always complete. For instance, design cooling-water outlet temperatures may be omitted, or the set of operating conditions may not, in fact, correspond to the stated design cleanliness factor. It is, therefore, good practice to verify design data sheets before using them for other purposes.

Tables 2.2 and 2.3 contain design data and calculations for four different condensers. Table 2.2 contains the core design performance data. This data is used in the set of fundamental calculations whose results make up lines 1 through 10 in Table 2.3. Lines 11 through 14 in Table 2.3 are the results from calculations based on the HEI method, while items 15 thru 21 are the results from using the ASME or thermal resistance method. The following notes outline the details of each calculation item.

Fundamental Calculations

1. The total tube surface area available for heat transfer is calculated from:

$$A = \pi \left(\frac{d_o}{12} \right) LN \tag{2.16}$$

Table 2.3. Calculation Results

	Material				
Quantity	Admiralty	Admiralty	70-30 Cu-Ni	AL-6X	Units
1 Surface area	27,763	55,417	50,411	172,507	ft^2
2 Flow area	8.963	15.65	25.65	65.21	ft^2
3 Density	62.32	62.20	62.37	62.26	lb_m/ft^3
4 Specific heat	0.9985	0.9980	0.9989	0.9982	
5 Mass flow	14,515	27,479	42,579	124,624	klb_m/h
6 Tube velocity	7.218	7.84	7.39	8.52	ft/s
7 Shell temperature	96.42	107.39	91.69	101.09	°F
8 LMTD	12.58	12.36	16.05	14.46	°F
9 Duty	214.93	452.51	433.81	1455.45	MBtu/h
10 Effective U coefficient	615.06	660.57	536.01	583.53	$Btu/(ft^2 \cdot h \cdot °F)$
11 UBASE	703.91	733.52	717.83	768.41	$Btu/(ft^2 \cdot h \cdot °F)$
12 Temperature correction factor	1.0244	1.0602	1.0002	1.0449	
13 U_{HEI}	721.08	777.64	631.82	682.46	$Btu/(ft^2 \cdot h \cdot °F)$
14 Cleanliness factor	85.29	84.95	84.83	85.5	%
15 Tube resistance	0.06139	0.06139	0.24059	0.30523	$°F/[Btu/(ft^2 \cdot h)]^*$
16 Water resistance	0.68860	0.59723	0.70401	0.55501	$°F/[Btu/(ft^2 \cdot h)]^*$
17 Shell resistance LH	0.42948	0.41589	0.42673	0.40043	$°F/[Btu/(ft^2 \cdot h)]^*$
18 Shell resistance MF	0.38723	0.36654	0.37793	0.40646	$°F/[Btu/(ft^2 \cdot h)]^*$
19 ASME U coefficient LH	847.83	930.65	729.22	793.23	$Btu/(ft^2 \cdot h \cdot °F)$
20 ASME U coefficient MF	879.33	975.45	756.13	789.45	$Btu/(ft^2 \cdot h \cdot °F)$
21 Performance factor	72.54	70.98	73.50	73.56	%

*All thermal resistances should be multiplied by 10^{-3}.

2. The flow area A_{fl} available to the cooling water is a function of not only the number of tubes and their inside diameter but also the number of passes and can be calculated from:

$$A_{fl} = \left(\frac{\pi}{4.0}\right)\left(\frac{d_i}{12}\right)^2\left(\frac{N}{N_{pass}}\right) \qquad (2.17)$$

3. The water density ρ can be computed from water temperature using the polynomial included in Table 1.1.
4. The specific heat of the water (C_p) can be computed from water temperature using the corresponding polynomial also included in Table 1.1.
5. The cooling-water flow is normally stated in GPM, but calculations of duty require the volumetric flow to be converted to mass flow W, in

lb_m/h; thus:

$$W = G \times 8.34 \times 60.0 \times \frac{\rho}{62.3} \qquad (2.18)$$

6. Tube water velocity can be calculated from volumetric flow and flow area:

$$V = \frac{G \times 8.34}{60 \times 62.3 A_{fl}} \qquad (2.19)$$

7. The ASME *Steam Tables* [1993] can be used to estimate the shell temperature corresponding to the design back pressure.
8. The log mean temperature difference (LMTD) can be calculated from the shell temperature T_v and the water inlet (T_{in}) and outlet (T_{out}) temperatures from:

$$\text{LMTD} = \frac{T_{out} - T_{in}}{\ln\left(\dfrac{T_v - T_{in}}{T_v - T_{out}}\right)} \qquad (2.20)$$

9. Duty can be calculated from:

$$Q = W C_p (T_{out} - T_{in}) \qquad (2.21)$$

10. The effective heat transfer coefficient based on the condenser operating conditions can be calculated from:

$$U_{eff} = \frac{Q}{A \times \text{LMTD}} \qquad (2.22)$$

HEI Calculations

11. UBASE can be calculated from tube water velocity V using Equation (2.4) or (2.5).
12. The HEI temperature correction factor F_W can be calculated using the appropriate equation of the set defined in Section 2.3.2 as a function of cooling-water inlet temperature.
13. The material correction factor F_M can be selected from the appropriate column and row of Table 2.1 and is included as an entry in Table 2.2. The expected value of the HEI overall tube-bundle heat transfer coefficient is then calculated from:

$$U_{HEI} = \text{UBASE} \times F_W F_M \qquad (2.23)$$

14. The design cleanliness factor DCF stated in the design data sheet may now be verified from:

$$DCF = 100.0 \frac{U_{\text{eff}}}{U_{\text{HEI}}} \tag{2.24}$$

Thermal Resistance Calculations

15. The tube wall resistance R_w can be calculated from Equation (1.11) of Chapter 1.

16. The water-film resistance can be calculated from Equation (1.14), remembering that water viscosity, density, and specific heat should be based on bulk water temperature defined as the mean of the water inlet and outlet temperatures.

17. One form of the shell-side resistance (LH) can be calculated from the Nusselt factor of Equation (1.1), the latent heat to be removed by condensation being the difference between the exhaust enthalpy on the expansion line for the load at condenser back pressure, minus the liquid enthalpy at the same back pressure.

18. Alternatively, the shell-side film resistance can be calculated as a function of exhaust mass flow rate (MF) using the Nusselt transformation contained in Equation (1.6) in Chapter1.0
Note that for both lines 17 and 18, the temperature difference/resistance sum ratios defined in Equation (1.16) must be used to compute the value of T_{wo}, or temperature of the outer tube surface

19. The ASME single-tube U coefficient for a clean condenser using the value of steam-side resistance resulting from step 17 can be calculated from Equation (1.15) of Chapter 1, with $R_{foul} = 0$.

20. The ASME single-tube U coefficient for a clean condenser using the value of steam-side resistance resulting from step 18 can also be calculated from Equation (1.15) of Chapter 1, with $R_{\text{foul}} = 0$.

21. The performance factor PF is the ratio of U_{eff} to the ASME single-tube heat transfer coefficient, expressed as a percent:

$$PF = 100.0 \frac{U_{\text{eff}}}{U_{\text{ASME}}} \tag{2.25}$$

Note that, since there is good data on the single-tube U coefficient for new tubes (see Section 1.5 and table 1.2), these having been calculated in accordance with item 19, this is the value of U_{SME} to be used in Equation (2.25).

Chapter 3

CONDENSER
PERFORMANCE MODELING

3.1 PERFORMANCE MODELING OBJECTIVES

The performance of the condenser can have a significant impact on the heat rate of a boiler/turbogenerator unit or on its generation capacity, both of which can be converted into cost terms. However, some economic impact may be a natural response to changing operating conditions, such as seasonal changes in circulating-water inlet temperature. In this case, circumstances may present an appropriate element of temperature control; in winter, for example, a cooling tower could be bypassed, and/or water flow could be reduced, possibly leading to economic benefit by reducing unnecessary condensate subcooling. Other sources of economic impact can be an increase in fouling deposition or leakage of air into the condenser. To correct either of these is a maintenance decision; and whether or when to take action will depend on the cost of that action compared with the losses incurred for the present and for some time into the future. The economics are also affected by whether the unit is operated in a base-loaded or a cycling mode.

Thus, a detailed knowledge of the economic effects is central to maintenance decision making, e.g., whether to clean and/or to isolate a leak. These effects must be quantified if informed decisions are to be made. It will be shown in Chapters 4 and 5 that the solution depends on solving sets of nonlinear simultaneous heat transfer/mass balance equations, the number and interconnections between which depend principally on the physical configuration of the condenser under study.

The manner in which the unit operates is also a factor when defining the structure of the equations. For example, the governors on steam turbogenerators in nuclear plants operate in a *boiler-follow* mode to maintain a desired steam pressure at the inlet to the throttle valve. The governors in some fossil fuel plants, particularly those on the steam turbines in combined-cycle plants, can also be made to operate in a boiler-follow mode. In all these cases, the amount of power generated is principally a function of the throttle flow, but it will also be positively affected by any drop in back pressure associated with improved condenser performance. Clearly, because throttle flow and exhaust flow are intimately related, it is the exhaust flow (but not its enthalpy) which is included as a given, or boundary condition, in the equation structure.

Alternatively, where the units are designed to operate in a *turbine-follow* mode, it is the boiler control system which regulates the fuel firing rate to maintain the desired steam pressure at the inlet to the throttle valve, the governor having been set to generate the desired amount of power. In this mode, the amount of power is a given; and it is the heat rate which will be affected by any reductions in back pressure associated with improved condenser performance.

Meanwhile, the duty on a condenser is primarily a function of the exhaust steam flow as well as its enthalpy. The exhaust flow, as the foregoing suggests, is either a given or a function of the amount of power being generated. The exhaust flow will also be affected by the enthalpy drop, which in turn is a function of condenser back pressure. Determining the exhaust enthalpy itself is complicated. The locus of the changing properties of the steam as it passes through the intermediate pressure (IP) and low pressure (LP) stages of the turbine is conventionally plotted on an enthalpy versus entropy graph, as included within the ASME *Steam Tables* [1993]. The coordinates of this expansion line vary with load, moving closer to the saturation line as the load falls. The enthalpy and entropy properties of the exhaust will also depend on the coordinates of the intersection of the back-pressure line with the expansion line corresponding to the current load. The interdependency of load, exhaust flow, and exhaust enthalpy requires that the condenser and the low-pressure stage of the turbine be considered as one interactive subsystem, the model of which will include an analysis of the steam properties along the expansion line corresponding to the current load.

With these considerations in mind, we designed the family of condenser performance models outlined in Chapters 4 and 5 to achieve the following set of objectives:

1. Estimate condenser duty as a function of load and back pressure. The latent heat to be removed from the exhaust of any steam-driven auxiliaries—e.g., boiler feedwater pump turbines—must also be included.

2. From this, estimate the cooling-water flow rate as a function of duty and cooling-water temperature rise. This estimate will reflect any reduction in flow caused by the fouling but will only indirectly be affected by any changes in heat transfer rates due to condenser fouling.

3. The condenser/LP stage model should be able to accommodate any condenser/LP stage configuration, including variations in the water flow path and the number of condenser compartments.

4. One function of the condenser/LP stage model is to calculate the duty if the condenser were operating under the same conditions but with the tubes clean. The losses due to the fouling can be calculated by subtracting the duty of the clean condenser from that of the fouled condenser.

5. A second function of this model is to estimate the fouling resistance in each compartment, as well as the water temperatures between compartments when they are connected in series (See Figures 3.3 thru 3.7).

6. If the water flow rate is less than the design value, and this is known to have been caused by tube or tubesheet obstruction, calculate the duty on the fouled condenser if the water flow were to be restored to its design value by cleaning or by the removal of the obstructions.

7. Organize the results calculated in the previous steps together with an

analysis of historical data to provide a database for optimizing the condenser cleaning schedule.

3.2 CONDENSER CONFIGURATIONS ENCOUNTERED IN PRACTICE

Figures 3.1 through 3.7 show a number of different condenser configurations which have been accommodated by the condenser/LP stage model. Given this variety, it is difficult to conceive of a configuration which could not be adapted to this model structure. Figure 3.1 shows a single-compartment condenser of the once-through type, or, alternatively, defined as having a single pass. A single-compartment condenser of the two-pass type is shown in Figure 3.2. These are the condenser configurations commonly provided for units generating less than 100 MW, marine condensers, and condensers associated with combined-cycle plants.

Larger units are occasionally provided with two single-pass condensers arranged in parallel and operating at the same back pressure. It is, however, more common for large units to be provided with condensers having two compartments, each operating at a different back pressure. Figure 3.3 shows one configuration in which continuous tubes are run through both compartments. It is customary to assume that the LP exhaust flow is evenly divided between the two compartments, because the pressure drop across both LP stages is sensibly the same. Because the water temperature in the second compartment is generally higher than that in the first, the back pressure in compartment 2 will be higher than that in compartment 1. Unfortunately, when the tubes run continuously through both compartments with no center waterbox provided, a mean intercompartmental water temperature cannot be measured, so that it is quite

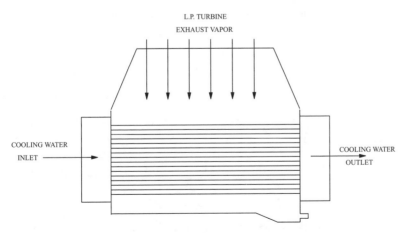

FIGURE 3.1. Single-compartment condenser, single-pass.

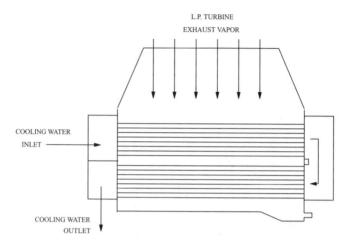

FIGURE 3.2. Single-compartment condenser, two-pass.

difficult to estimate the performance of each compartment individually. However, the condenser/LP stage model can be configured to estimate this temperature, from which duties and fouling resistances can also be estimated.

Figure 3.4 shows the configuration when two separate condensers are provided but the water path flows in series from one compartment to the other. Again, because of the generally higher water temperatures in compartment 2 than compartment 1, the back pressure in compartment 2 is also higher. However, the turnaround provides the opportunity to measure the intercompartmen-

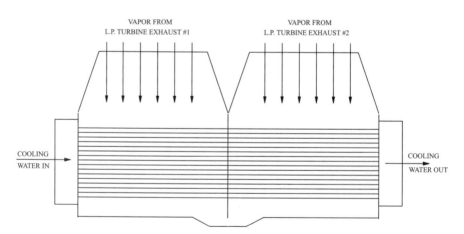

FIGURE 3.3. Two compartments with different back pressures-continuous tubes.

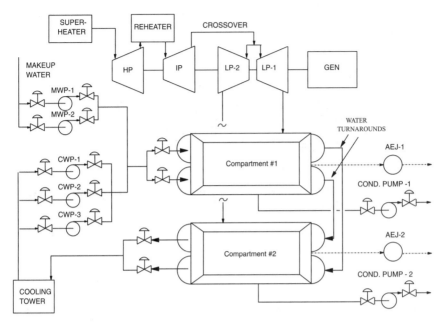

FIGURE 3.4. Two compartments with different back pressures, with turnaround between compartments.

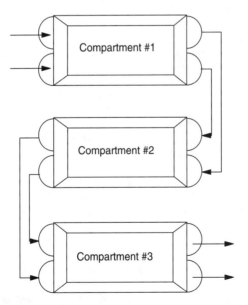

FIGURE 3.5. Three compartments with different back pressures, with turnaround between compartments.

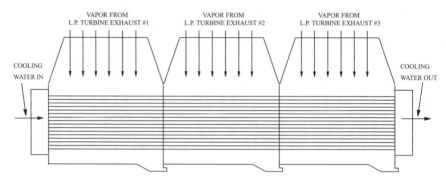

FIGURE 3.6. Three compartments with different back pressures, with continuous tubes.

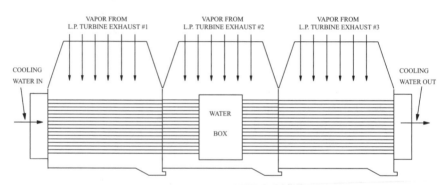

FIGURE 3.7. Three compartments with different back pressures, with waterbox in center compartment.

tal water temperature directly, although, for some reason, in very few plants have these sensors been provided as a matter of course. Figure 3.4 also shows some of the pumping equipment which is normally furnished as a part of the condenser subsystem.

Larger units, as well as many nuclear power plant units, are often provided with three-compartment condensers. Each of the three LP stages is located immediately above the associated compartment, while the cross-sectional area of the crossover connecting the IP stage to the three LP stages is reduced after each stage to ensure that the total LP exhaust flow is evenly distributed among the three stages.

Figure 3.5 shows the configuration when three separate condenser compartments are provided with the cooling water flowing through them in series. Clearly, this configuration will allow the intercompartment water temperatures to be measured directly, thus facilitating the performance analysis of each individual compartment. Figure 3.6 shows another arrangement, where the tubes run continuously from one compartment to the next. This, again, precludes direct measurement of intercompartmental water temperatures, so that more reliance has to be placed on the models when computing the performance of each compartment. Figure 3.7 shows another configuration for a three-compartment condenser, provided for a fossil fuel plant, and having an additional waterbox installed in the center compartment. This permits an intermediate water temperature to be measured and improves the confidence in the results, but unfortunately this does not represent either of the intercompartmental water temperatures directly.

It is clear from this variety of condenser configurations that the condenser/LP stage model must be structured to accommodate the configuration found at a particular site, but without distorting the fundamental set of heat and mass balance equations used throughout the model calculations.

3.3 INSTRUMENTATION

Section 4.0 of PTC.12.2-1998 [ASME 1998] provides a comprehensive overview of the type and location of instrumentation required to perform a condenser acceptance test or performance test. However, many of the recommendations are not suitable for continuous monitoring of condenser performance.

3.3.1 Condenser Pressure

Condenser pressure should be measured at a point in the condenser between 1 ft and 3 ft above the tube bundle. Ideally, pressure should be measured at three points in a shell, and a set of three points should be provided for each separate compartment. Basket tips or guide plates should be installed to reduce the kinetic effect of vapor velocity and allow the static pressure to be measured. All piping and connections must be both steam- and airtight; and the piping must be arranged so that it falls continuously, allowing condensate to be continuously drained from the tubing which usually connects the pressure transmitters to the condenser. Pockets of condensate must not be allowed to accumulate; otherwise, they will affect the accuracy and stability of the pressure measurements. It is recommended that electronic absolute-pressure transducers be used, having a maximum uncertainty of ±0.01 in-Hg.

3.3.2 Cooling-Water Inlet Temperature

Because the cooling water at the condenser inlet is well mixed, only one temperature measurement device is usually necessary. However, where compartments are arranged in parallel, a small difference between the waterbox inlet temperature measurements to the two compartments has sometimes been observed. When this occurs, it should be carefully explored, first by ensuring that the temperature sensors are properly calibrated and then by investigating the reason for the discrepancy. Temperature sensors should have an accuracy of ±0.1°F, and Resistance Temperature Devices (RTDs) or thermistors are recommended.

3.3.3 Cooling-Water Outlet Temperatures

Cooling-water outlet temperatures are more difficult to measure, because of the large variations caused by different heat transfer coefficients throughout the tube bundle. Stratification can also occur in the outlet piping, so that truly representative temperature values can be obtained only after the water leaving the condenser has been thoroughly mixed.

Where a condenser is provided with several outlet waterboxes, it is quite common for the temperature in each to be measured separately. Clearly, the location of the sensor will be important in determining how closely it is measuring a *mean* outlet water temperature for the respective waterbox. Differences in waterbox temperatures should be carefully analyzed. After verifying sensor calibration, it may be found that the differences are caused by flow obstructions. Asymmetrical discharge of the exhaust from auxiliary turbines or leaking bypass valves can be other causes of temperature variations. The mere averaging of the outlet temperatures may not provide a value which truly reflects the rate and distribution of condensation taking place within the shell.

3.3.4 Cooling-Water Flow Rate

Cooling-water flow rate is a vital part of condenser performance calculations. Not only is it used to calculate tube water velocity, from which the HEI base heat transfer coefficient is determined (see Section 2.3.1), but it is also the most significant parameter in the tube-side resistance value calculated in Equation (1.14) from the Rabas/Cane correlation. However, it is seldom measured directly in practice, although PTC.12.2-1998 [ASME 1998] describes the following flow measurement methods as being acceptable for performing condenser acceptance tests:

1. Velocity traverse
2. Tracer dilution
3. Differential producer
4. Ultrasonic time of travel
5. Energy balance

The first two methods are not very suitable for continuous performance monitoring. Some means of creating a differential pressure would be appropriate for this purpose, but the piping is usually so large that the cost of installing a flowmeter using orifice plates or annubars is normally considered prohibitive. Ultrasonic time of travel testing is also a valid technique, but the cost is again high. Energy-balance methods are often performed where units have been equipped with a comprehensive data-acquisition system. However, establishing the amount of energy rejected to the condenser may involve performing calculations using some 70 different pressure, temperature, and flow transducers, and the accuracy of the calculated net energy to be removed by the condenser may not be adequate for obtaining a meaningful and relatively noise-free cooling-water flow rate.

It is for these reasons that the models described in Chapter 4 use an indirect method based on developing a good estimate of the condenser duty from an analysis of thermal kit data, generated power, and condenser back pressure. The flow rate is then obtained by dividing the duty by the water temperature rise. Any noise contained in this value can be attenuated by means of digital filters, and some of the plots obtained using this method are included in Chapter 6.

3.3.5 Waterbox Levels

It is very important that both inlet and outlet waterboxes are always running full, and means should be provided to monitor the levels in them. If an inlet waterbox is not running completely full, then some of the upper tubes will be uncovered and so unable to participate in the heat exchange process. This occurs if there is insufficient circulating-water pump discharge head. If the outlet waterboxes are not running full, the flow through the tubes will also be affected [Fromberg 1997]. Air leaks into an outlet waterbox can cause the siphon to be broken. Not only will this reduce the pressure drop across the tubes, but it can, again, cause some of the upper tubes to become uncovered.

3.3.6 Waterbox Differential Pressures

Measurement of waterbox differential pressure can provide a measure of the water flow rate if the tubes are clean. In addition, the pressure drop across the waterboxes when compared with a measure of the flow rate can be an indicator of the degree of fouling which has occurred. However, waterbox differential pressure can be affected by the presence of air, especially in the outlet waterboxes; or by air leaks which cause the outlet waterboxes not to run full.

3.3.7 Tube-Bundle Fouling

PTC.12.2-1998 [ASME 1998] mandates an interesting method for determining tube-side fouling resistance. Pairs of tubes are selected, the minimum number of pairs required being the number of tubes in the tube bundle divided by 2000. However, there should be no fewer than 4 pairs and not more than 16 pairs per bundle. One tube of the pair, the fouled tube, is provided with temperature sensors at the inlet and outlet. The other tube is a reference tube and either is new or has just been cleaned. The water outlet temperature on this tube is also measured. The assumption is that the two tubes experience the same heat transfer conditions of vapor pressure and cooling-water velocity. The fouling resistance at tube pair i can be calculated from:

$$R*_{f_i} = \frac{A}{WC_p}\left[\frac{1}{\ln\left(\dfrac{T_v - T_{1,f_i}}{T_v - T_{2,f_i}}\right)} - \frac{1}{\ln\left(\dfrac{T_v - T_{1,c_i}}{T_v - T_{2,c_i}}\right)}\right] \tag{3.1}$$

and

$$R*_f = \frac{1}{j}\sum_{i=1}^{j} R*_{f_i} \tag{3.2}$$

A major problem with this method is that, while the relative cleanliness of the two tubes in a pair will be retained during the period of an acceptance test, both tubes will subsequently tend to foul at the same rate, so that the second tube can no longer be regarded as a reference.

Meanwhile, EPRI and Bridger Scientific [Garey 1996] have developed another, but somewhat expensive, method for monitoring tube fouling in which the state of the reference tube remains unchanged over time. Again, pairs of tubes are assigned throughout the tube bundle. The temperatures at both the inlet and the outlet of the active tube are measured as well as the flow of water through it. However, in this case, the reference tube is plugged at both ends. It is also provided with temperature sensors, measuring the temperature T_v inside the reference tube, which, under steady-state conditions, will assume the same value as the shell temperature at its location in the tube bundle. The heat transfer coefficient of the active tube is calculated from:

$$U = \frac{WC_p}{A}\ln\left(\frac{T_v - T_{in}}{T_v - T_{out}}\right) \tag{3.3}$$

and may be plotted to monitor changes in tube condition. The fouling resistance may be calculated by comparing the present value of the heat transfer coefficient (U_{eff}) with the highest value achieved after the latest tube cleaning (U_{ref}):

$$R_f = \frac{1}{U_{eff}} - \frac{1}{U_{ref}} \tag{3.4}$$

Alternatively, the fouling resistance can be calculated using the steam-side film ($1/h_f$), tube-wall (R_t), and water-side film (R_w) thermal resistances calculated in accordance with Equations (1.1), (1.11), and (1.14), from which:

$$U_{ref} = \frac{1}{\dfrac{1}{h_f} + R_w + R_f} \tag{3.5}$$

Chapter 4

MODEL OF TUBRINE
LOW-PRESSURE STAGE
AND ESTIMATION OF
CONDENSER DUTY

4.1 INTRODUCTION

In the utility industry, the steam generators (fossil fuel or nuclear), steam turbines, and condenser are configured as a closed system to operate in a sequence known as the Rankine cycle. Water is the energy conversion medium, existing in either liquid or vapor forms, depending on the point in the cycle. Vendors of steam turbines provide extensive computer printouts (frequently referred to as thermal kits) showing the steam and water conditions at every significant point in the cycle and for various loads, commonly 100%, 75%, 50%, and 25%, although other loads and sets of operating conditions are often also included. It is common to plot the conditions throughout the Rankine cycle on a Mollier diagram for steam and water. In the case of steam expansion through the low-pressure stage of the turbine, it was indicated in Chapter 3 that the position of the expansion line on the Mollier diagram shifts with load, and this affects the amount of heat which has to be removed from the vapor entering the condenser, as well as the amount of thermal heat which is converted to power generation.

The thermal kit data published by the turbine vendor is carefully verified during turbine acceptance tests and does not normally change significantly over time. Even if rotors with more efficient blading are subsequently installed, it is customary to update the thermal kit data to reflect the improved performance. Thus, thermal kit data can be used as a benchmark for estimating the amount of heat in the vapor which has to be removed by the condenser, this depending on the amount of power being generated as well as the back pressure(s) in the condenser compartment(s).

4.2 THE STATES AND PROPERTIES OF THE WATER
MOLECULE

The water molecule can exist in the solid, liquid, or gaseous state. The ASME *Steam Tables* [1993] show the properties of the water molecule plotted in terms of

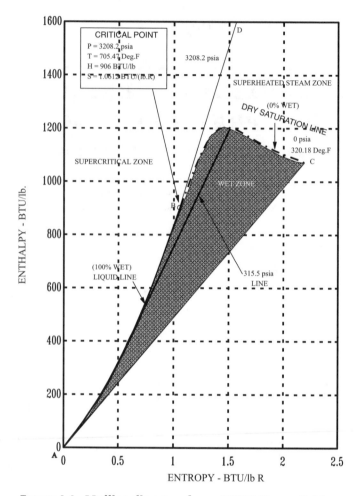

FIGURE 4.1. Mollier diagram from ASME Steam Tables.

the enthalpy/entropy relationships, and these are summarized in the plot shown in Figure 4.1. In the solid state, water is frozen; it becomes liquid after receiving *latent heat of fusion*. To change water from the liquid to the gaseous state while it is held at a constant pressure and temperature, *latent heat of vaporization* has to be added.

A study of Tables 1 and 2 of the ASME *Steam Tables* will show that, for any temperature below the critical temperature of 705.47°F, when water is in equilibrium with its vapor in a pressure vessel, there is a corresponding saturation pressure and vice versa. If the vapor has acquired exactly the amount of latent heat required to bring it to a dry gaseous state, it is termed *dry saturated* and its moisture content is deemed to be 0%, or, alternatively, the vapor is said to be 100% dry. When in the liquid state or condensed, vapor may be considered

100% wet. Any water in contact with the vapor in such a two-phase system will assume the enthalpy and entropy of condensed liquid at the appropriate pressure and temperature.

The central feature of the Mollier diagram for water and steam, plotted with enthalpy and entropy coordinates, is the area bounded by lines *AB*, *BC*, and *CA* of Figure 4.1. Line *AB* is a plot of enthalpy and entropy of the water molecule in the liquid state, i.e., condensed liquid containing no latent heat of vaporization; this locus is known as the *liquid line*. The pressure along line *AB* varies between 0 (point *A*) and the critical pressure of 3208.2 psia (point *B*), the temperature of the water being the saturation temperature corresponding to the pressure. At low pressures, these are the conditions of the condensate in the hot well of a condenser.

Point *A* of Figure 4.1 (32.018°F) shows the enthalpy and entropy of the water molecule in the liquid state to be both essentially zero, as is also the saturation pressure corresponding to that temperature. If latent heat of fusion is removed from liquid water at this temperature, the molecule will assume a solid (frozen) state.

Line *BC*, known as *the saturation line*, is a plot of the enthalpy/entropy relationship of the water molecule in a dry saturated gaseous state, for pressures between the critical point of 3208.2 psia (point *B*) and 0 psia (point *C*). The moisture content of the vapor on or above this line is 0%, and the vapor is termed 100% dry. At pressures below the critical value of 3208.2 psia but at temperatures *above* the saturation value corresponding to the pressure, i.e., above line *BC*, the vapor is said to be *superheated*.

Line *CA* connects the enthalpy/entropy coordinates of dry saturated vapor at 0 psia with those for condensed water at the same pressure. Within the area *ABCA*, defined by the points *A*, *B*, and *C*, the vapor is somewhere between the dry saturated and liquid states and is said to be *moist or wet steam*. By drawing a straight line between the enthalpy/entropy coordinates for the liquid (100% wet) and dry saturated (0% wet) states for a given pressure or temperature, the properties of the vapor at varying degrees of wetness between 0% and 100% can be estimated, by triangulation along this line. The vapor entering a condenser from a low-pressure turbine exhaust is often wet.

In Figure 4.1, point *B* is the *critical point*, at which the enthalpy and entropy of water in both the liquid and gaseous states are exactly the same so that it is uncertain in which of these two states a molecule actually is. At the critical point, the pressure is 3208.2 psia, the temperature is 705.47°F, the enthalpy 906.0 Btu/lb$_m$, and the entropy 1.0612 Btu/(lb$_m$ · °R). Note that the liquid and saturation lines have identical enthalpy/entropy coordinates at the critical point *B*. Further, at temperatures and pressures above those at the critical point, the water molecule can exist only as a vapor.

Also shown in Figure 4.1 is the enthalpy/entropy locus (*BD*) corresponding to the critical pressure of 3208.2 psia. The supercritical zone lies to the left of BD and, below point *B*, the supercritical zone can also be considered as lying to the left of the liquid line, for all practical purposes.

At pressures below the critical point, water can also exist at temperatures below the saturation temperature corresponding to the pressure, and it is then known as *compressed water*.

4.3 THE RANKINE CYCLE

The Rankine cycle is a closed circuit in which water in different states is the circulating medium. Heat is transferred to or extracted from the medium in order to convert heat from an energy source to useful work. In a fossil fuel–fired power plant, the Rankine cycle can be described as beginning with the action of a condenser, from which condensate is drawn. The condensate is then preheated and pumped under high pressure to a boiler, in which it is converted to steam at the desired pressure and temperature by the combustion of fuel. The steam is then expanded through several stages of a steam turbine to generate power, the exhaust from the last stage being condensed in the condenser, thus closing the cycle.

The Rankine cycle may also be plotted on a Mollier diagram in terms of the enthalpy versus entropy path followed as energy is imparted to or extracted from the liquid or vapor. A typical plot for a subcritical pressure reheat cycle is shown in Figure 4.2. Starting at the hot well (point E), the condensate will assume the conditions of water in the liquid state at the pressure of the LP turbine exhaust, in this case 2.5 in-Hg. The condensate is drawn from the condenser by condensate pumps and passes through several noncontact-type low-pressure feedwater heaters before entering the deaerator (point F), in which the temperature of the water will assume the saturation temperature corresponding to the pressure of the extraction steam being supplied to the deaerator.

The water is then drawn from the deaerator by high-pressure feedwater pumps, the discharge pressure from which is a function of, and higher than, the pressure at the turbine throttle (point H). After passing through a train of high-pressure feed heaters, as well as the economizer of the boiler, the water enters the steam drum located at the top of the boiler, the pressure in which is again higher than that at the throttle. Downcomers allow water to be drawn from the steam drum and then circulate within the water walls, through which latent heat is acquired from fuel combustion. Within the water walls some of the water is in the liquid state and some in the wet vapor state (i.e., the water exists in two phases). This mixture moves upward through the tubes forming the water walls, and as it reenters the drum, the vapor becomes separated from the water, the latter recirculating back through the downcomers to the walls.

Because of the pressure drop between the drum and the turbine throttle valve, the vapor, which is still slightly moist (point G), leaves the drum and passes through one or more superheaters. After acquiring the balance of the latent heat, the steam passing through the superheater rises in temperature above the saturation value corresponding to the pressure, until it reaches the required degree of superheat (point H). Note that there is a pressure difference between the vapor in the drum and the superheated steam leaving the boiler. The superheated steam then passes through the turbine throttle to the high-pressure stage of the turbine, through which it expands; by expanding, it generates an amount of power proportional to the drop in enthalpy between points H and I.

The exhaust from the high-pressure stage then reenters the boiler and passes through the reheater, in which the steam temperature is raised once more, at a sensibly constant pressure, until the temperature at the outlet of the reheater (point J). is close to the design value. After leaving the reheater, the steam is

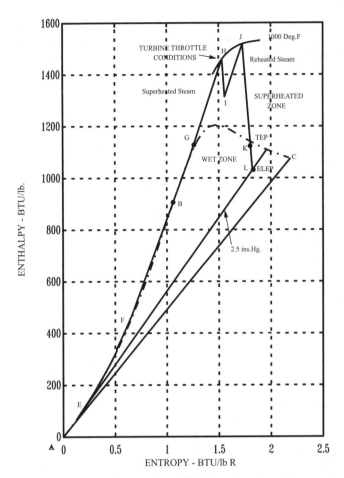

FIGURE 4.2. Rankine cycle diagram.

then expanded through the intermediate- and low-pressure stages of the turbine and again generates an additional amount of power proportional to the drop in enthalpy between point *J* and the turbine end point (TEP), *K*. Note that, as the vapor leaves the low-pressure stage at point *K*, it is usually slightly wet. The low-pressure (LP) exhaust then expands slightly more through the exhaust annulus down to *L*, the expansion line end point (ELEP), and finally passes to the condenser, within which all heat remaining in the exhaust vapor is removed by the cooling water circulating through the condenser tubes. The vapor is thus finally recovered as condensate, which can then be recycled through the system.

Although the Rankine cycle is essentially a closed system, some water is lost in the form of boiler blowdown, a continuous or intermittent process through which any inorganic salts or deposits are removed from the system. Some vapor leaks also occur, either in the form of steam leaking to the atmosphere through turbine shaft glands; or the vapor bled off from the deaerator; or the vapor removed from the condenser by the air removal equpment in an attempt to maintain the concentration of non-condensible gases within the system at an acceptable level. These water and vapor losses are replaced, usually by treated makeup water, which enters the system in the condenser hot well or at the deaerator.

4.4 THE LOW-PRESSURE STAGE EXPANSION LINE

Section 4.3 described the main features of the Rankine cycle for a particular load point. In fact, as the load changes, some of the points plotted on the Mollier diagram move. For instance, while boiler controls maintain the superheated steam conditions essentially constant, the coordinates of points I, J, and K will shift with load. It is well known that the turbogenerator heat rate increases as the load falls, and the associated falling off in efficiency results in an increase in enthalpy throughout the IP and LP stages. Figure 4.3 shows the shifting of the IP and LP stage expansion lines with respect to load; note that as the load falls, the entropy and enthalpy increase while the expansion line end point (ELEP) moves closer to the saturation line.

The turbine thermal kit data set defines the set of conditions which are to be expected at a given load, with the stated design back pressure having the same value for all loads. Thus, regardless of the load, all ELEPs will fall on the line corresponding to the design back pressure and connecting together the enthalpy/entropy coordinates for condensed liquid and dry saturated vapor at that pressure.

The other significant operating point is the *turbine end point (TEP)*, also known as the *used energy end point (UEEP)*, a typical locus for which is also shown in Figure 4.3. The enthalpy difference between the TEP locus and the corresponding ELEP is known as the *exhaust annulus loss*, a typical example of which is shown in Figure 4.4. The flat metal ring which surrounds the turbine shaft at the exit from the low-pressure stage of the turbine creates an annulus through which the vapor passes before it enters the condenser. However, under certain operating conditions, the vapor velocity can approach the value at which *choked flow* occurs (see Section 4.5.1). Thus a complete analysis of the low-pressure expansion line must take this possibility into account.

Note that the end of the low-pressure stage expansion line tends to lie within the wet zone of the Mollier diagram, which means that the vapor properties at any point on the expansion line can be determined only by calculating the vapor and liquid properties at the intersection of the expansion line and the line in the wet zone corresponding to the back pressure.

With nuclear power plants, or fossil fuel plants operated in the *boiler-follow mode*, the exhaust flow will be a given boundary condition, and improvements in back pressure will result in an increase in generated power. However, with fossil fuel plants operated in the *turbine-follow mode*, the amount of power generated will be a given boundary condition and improvements in condenser back pressure will result in conversion of more energy in the throttle steam to power, accompanied by a reduction in both throttle flow (under the action of the turbine

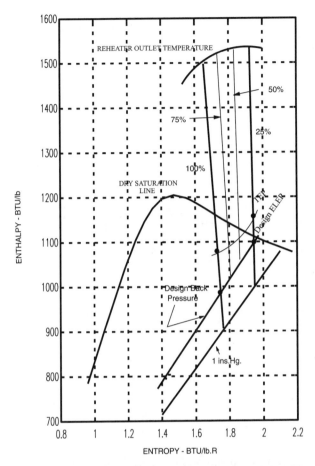

FIGURE 4.3. End-point expansion lines vs. load.

governor) and the associated exhaust flow. The drop in back pressure will also cause the condenser duty to decrease, assuming that the annulus is not operating under choked conditions.

4.5 ANNULUS EXHAUST LOSSES

Annulus exhaust losses (in Btu/lb$_m$) are normally plotted with respect to annulus velocity V, in feet per second. However, some vendors plot the losses with respect to annulus volumetric flow (VOLFLOW), usually in millions of cubic feet per hour, and it is convenient to convert such plots to the velocity equivalent, using the following relationship:

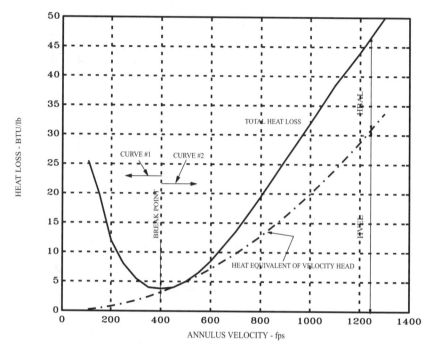

FIGURE 4.4. Annulus exhaust loss vs. annulus velocity.

$$V = \frac{\text{VOLFLOW}}{3600 \times \text{AREA}_{\text{ann}}} \tag{4.1}$$

where AREA_{ann} is the total annulus area, a parameter usually stated on the curve provided by the turbine manufacturer. The relationship between losses and velocity is needed to analyze the LP expansion line, but the curve typically has the shape shown in Figure 4.4, which is difficult to model if only one equation is to cover the whole flow range. For this reason, the curve should be split into two parts, separated at the velocity corresponding to the minimum annulus exhaust loss value (V_{min}). The coordinates of sets of points in the two zones are then abstracted and regressed to obtain the sets of regression coefficients for two different third-order polynomials each having the form:

$$H_{\text{ann}} = a_1 + a_2(V - V_{\text{min}}) + a_3(V - V_{\text{min}})^2 + a_4(V - V_{\text{min}})^3 \tag{4.2}$$

Figure 4.4 also includes a plot of the enthalpy equivalent of the velocity head, calculated from:

$$H_{\text{vel}} = \frac{V^2}{778 \times 64.4} \tag{4.3}$$

and it is seen that the annulus exhaust loss includes other factors. According to Robinson [1933], these are composed of:

- Losses associated with the exit velocity of the steam as it flows through the exhaust annulus
- Loss due to tangential velocity component-whirl loss
- Eddy losses associated with the nonuniform flow through the annulus
- Pressure drop through the exhaust hood

4.5.1   Choked Flow at the Annulus

With unchoked flow through the annulus, the vapor will expand down to an end-point pressure which is the same as the back pressure in the condenser shell, following the path *RST* shown in Figure 4.5. The turbine end point (TEP) will then be equal to the enthalpy on the expansion line which corresponds to the back pressure (ELEP), plus the annulus exhaust loss equivalent to the velocity of the exhaust vapor at the *inlet* to the annulus.

However, choking can occur when supersaturated steam passes through an annulus at high velocity. Lewitt [1953] defines choking as occurring when the ratio of pressure at the throat of the annulus to the inlet pressure (or pressure at the TEP) is equal to or less than 0.58. When this occurs, the maximum mass flow of vapor is limited to the flow corresponding to this pressure ratio; at the same time, the *enthalpy at the exit of the annulus* cannot fall below the enthalpy at the throat under choking conditions. Thus, the expansion follows path *RSTU* shown in Figure 4.6, expanding down to the throat pressure along the expansion line, but then expanding at constant enthalpy along line *TU* down to the back pressure currently existing in the condenser shell. Thus, under choked conditions, reducing the back pressure below the critical value will neither improve the exhaust enthalpy (i.e., reduce the condenser duty) nor reduce the amount of steam or throttle flow required to sustain a given load.

4.6   LOW-PRESSURE EXPANSION LINE WITH CURVATURE

A paper by Spencer, Cotton, and Cannon [Spencer 1974] contains equations for plotting expansion lines based only on the enthalpy/entropy coordinates of the steam conditions at the inlet and outlet of a stage. The equations on page 51 of the original paper are as follows. Let

h_A, s_A = enthalpy/entropy coordinates of expansion line at stage inlet
h_B, s_B = enthalpy/entropy coordinates for expansion line at stage outlet
h = enthalpy at any point along expansion line
s = corresponding value of entropy

For the high-pressure and intermediate sections,

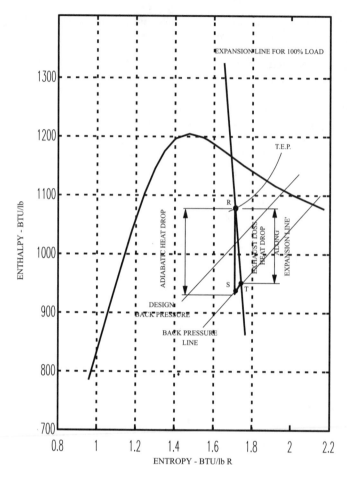

FIGURE 4.5. Expansion under normal conditions.

$$s = s_B + (s_A - s_B) \frac{h - h_B}{h_A - h_B} \quad (4.4)$$

For the reheat and non-reheat sections,

$$s = 10^{[h_B - (h + H_0)]/371.0} + R_0(h - h_B) + s_B - k_0 \quad (4.5)$$

and

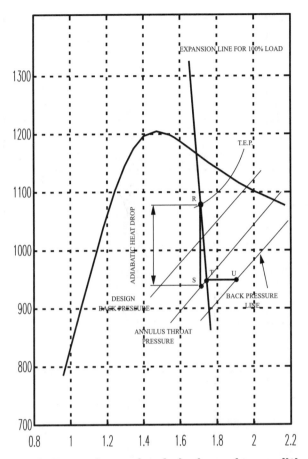

FIGURE 4.6. Expansion under choked annulus conditions.

$$R_0 = \frac{(s_A - s_B) + k_0 - 10^{[h_B - (h_A + H_0)]/371.0}}{h_A - h_B} \tag{4.6}$$

For reheat sections,

$$H_0 = 650.0 \qquad k_0 = 0.0177$$

For non-reheat sections,

$$H_0 = 450.0 \qquad k_0 = 0.06124$$

Note that there is an error in the original paper, which assumes that k_0 in Equa-

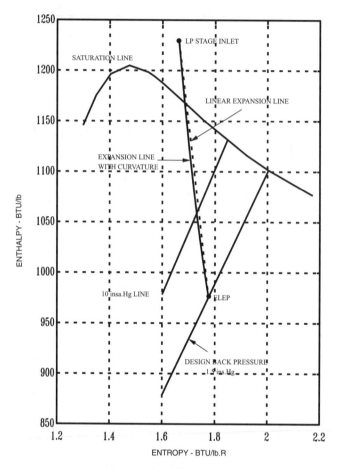

FIGURE 4.7. Expansion line with curvature.

tions (4.5) and (4.6) is assigned the value of 0.0177 whatever the value of H_0. However, if the curve is to pass through end points (h_A, s_A), (h_B, s_B), and $h = h_B$, then s can be equal to s_B only if:

$$k_0 = 10^{-H_0/371} \qquad (4.7)$$

Figure 4.7 compares the expansion line plot assuming a linear relationship between enthalpy and entropy throughout the low-pressure stage and the equivalent plot with curvature, using Equations (4.5) and (4.6). From the range of 10 in-Hg down to, in this case, the design back pressure of 1.5 in-Hg, there is only a small difference in the slopes of these plots at the ELEP. George J. Silvestri, Jr. [1997], has termed the linear plot a "useful fiction," and, since the small difference in slope has very little effect on the significance of any calculations, the simplification introduced by the linearity justifies a linear representation of the low-pressure expansion line.

4.7 EXPANSION LINE ANALYSIS

Much useful information about the behavior of a condenser can be gained from analysis of the vapor properties along the LP stage expansion line of the associated condensing steam turbine. Figure 4.3 shows how the location of a plot of the low-pressure expansion line on the Mollier diagram varies with the load. For a given load, the enthalpy/entropy (H/S) coordinates at the entry to the IP or LP stage can be found from the turbine thermal kit data, which also includes the design enthalpy at the corresponding expansion line end point (ELEP), as well as the design back pressure. The thermal kit data also contains the LP stage exhaust flow, and usually the design turbine end point (TEP) corresponding to the load.

The condenser duty is a function of the product of exhaust enthalpy and flow rate, but in the *turbine-follow* mode, both will be affected by any deviations from condenser design back pressure. Expansion line analysis will allow calculation of the exhaust enthalpy, together with the turbine end point, for any back pressure. However, in this operating mode, variations in back pressure and associated ELEP will temporarily vary the amount of energy converted to mechanical work, i.e., the amount of generated power. The turbine governor, sensing this, will adjust the throttle flow to maintain the desired power level and, in so doing, will have a consequent effect on the exhaust flow rate. At equilibrium, the associated adjustment to design exhaust flow can be approximated by multiplying the design exhaust flow corresponding to the load by the ratio of the enthalpy difference between throttle and operating UEEP to the enthalpy difference between throttle and design UEEP. See Equation (4.14), later in this chapter.

With a unit operating in the *boiler-follow* mode, the exhaust flow is the given and, ideally, the location of the expansion line should be defined in terms of exhaust flow rather than generated power. In this mode, variations in condenser back pressure about the design value will not affect the exhaust flow rate but will vary the UEEP, together with the power generated from the throttle and exhaust flows.

In either operating mode, the unit may be susceptible to choking of the annulus under high exhaust flow and low back-pressure conditions. Expansion line analysis should thus include determination of the maximum exhaust flow and associated annulus throat pressure for a given load, both being potential constraints on a turbine/condenser model, as described in Chapter 5.

A simplistic understanding of the usefulness of expansion line analysis might suggest compiling steam table data in a complicated lookup table for later reference. In practice, however, not only does the expansion line analysis software follow the fundamental steam property relationships more faithfully: it can always be executed in real time *for a given load* as an integral part of condenser performance evaluation.

The expansion line analysis program requires that several relationships first be developed from the set of thermal kit data supplied by the turbine vendor. Defined in terms of the set of regression coefficients which provide a minimum least-squares fit for a third-order polynomial, the following is the set of relationships to be regressed for a turbine operating in the turbine-follow mode:

Design Data Polynomials

1. Turbine exhaust flow versus megawatt load
2. Exhaust enthalpy at design back pressure (ELEP) versus megawatt load
3. Turbine end-point enthalpy or useful energy end point (UEEP) versus megawatt load
4. Design reheater outlet pressure versus megawatt load
5. Exhaust loss versus annulus velocity—velocities below that corresponding to minimum loss
6. Exhaust loss versus annulus velocity—velocities above that corresponding to minimum loss

The data for polynomials 1 through 4 will be found in the mass/energy balances contained in the thermal kit, while the data for items 5 and 6 can be abstracted from the annulus exhaust loss curve, also usually included within the thermal kit.

4.8 ORGANIZATION OF THE EXPANSION LINE ANALYSIS PROGRAM

Before discussing the organization of the expansion line analysis program, let us consider the way the results will be used in the turbine/condenser model described in Chapter 5. Table A.1 in Appendix A shows a typical set of variables for a single-compartment condenser, one of which is shell temperature (designated STMP1). This, rather than back pressure, has been chosen because the shell-side vapor temperature is itself a significant parameter in the calculation of log-mean temperature difference, which is embedded in the basic Fourier heat transfer equation, Equation (1.21). Thus, the results from an expansion line analysis should be presented *in relation to the temperature of the vapor in the condenser shell.*

Unfortunately, the annulus loss is calculated from the annulus velocity of the vapor, and this depends on the pressure at the turbine end point, not condenser back pressure. However, by scanning the expansion line over a range of turbine end-point pressures, calculating a number of vapor properties at each point, and regressing the results in an appropriate sequence, it is possible to obtain sets of regression coefficients for UEEP and ELEP in terms of shell-side vapor temperature, or shell temperature. To achieve this result, the sequence of tasks to be performed within the analysis of a low-pressure turbine stage expansion line is as follows:

1. Calculate H_{in} and S_{in}, the HS coordinates of the expansion line at the inlet to the IP or LP stage. The steam pressure at the stage inlet can be calculated using polynomial 4 of Section 4.7, and the associated design reheat steam temperature at that point will be a part of the database. The steam tables can now be used to obtain the enthalpy and entropy of superheated steam corresponding to this pressure and temperature.
2. Calculate the entropy coordinate S_{ELEP} corresponding to the enthalpy at the expansion line end point H_{ELEP}, the latter being obtained from polynomial 2. This is found by interpolating the design back-pressure line between the liquid and dry saturated vapor enthalpy and entropy values corresponding to that pressure. Let

$$H_{vap}, S_{vap} = HS \text{ coordinates of intersection of design}$$
$$\text{back-pressure line on dry saturation line}$$
$$H_{liq}, S_{liq} = HS \text{ coordinates of intersection of design}$$
$$\text{back-pressure line on liquid line}$$

Then

$$S_{ELEP} = S_{liq} + (S_{vap} - S_{liq})\frac{H_{ELEP} - H_{liq}}{H_{vap} - H_{liq}} \qquad (4.8)$$

3. Using the set of HS coordinates calculated for the two ends of the expansion line, calculate the coefficients in a first-order relationship between enthalpy and entropy. Let

$$H_{in}, S_{in} = HS \text{ coordinates of expansion line}$$
$$\text{at IP or LP stage inlet}$$
$$H_{ELEP}, S_{ELEP} = HS \text{ coordinates of expansion line}$$
$$\text{at ELEP}$$

The relationship between enthalpy and entropy along the expansion line is of the form:

$$H_{exp} = a_0 + a_1 S_{exp} \qquad (4.9)$$

where

$$a_1 = \frac{H_{in} - H_{ELEP}}{S_{in} - S_{ELEP}}$$
$$a_0 = H_{ELEP} + a_1 S_{ELEP}$$

4. Scan the expansion line in terms of turbine end point (TEP) pressure, starting at 10 in-Hg and decrementing in 1 in-Hg steps down to 1 in-

Hg. At each step, the steam tables are used to calculate the following set of vapor properties (or parameters) on the expansion line which correspond to the pressure at TEP:

a. Saturation temperature at TEP
b. Specific volume at dry vapor saturation line
c. Enthalpy at the dry vapor saturation line
d. Entropy at the dry vapor saturation line
e. Enthalpy at the liquid line
f. Entropy at the liquid line

Between 10 in-Hg and 1 in-Hg, the expansion line will almost certainly lie within the wet zone of the Mollier diagram. Thus, at each step, it is necessary to calculate the HS coordinates at the intersection of the expansion line with the line corresponding to the pressure and lying within the wet zone. If a linear relationship between enthalpy and entropy along the pressure line is assumed, it will be of the form:

$$H_{press} = b_0 + b_1 S_{press} \qquad (4.10)$$

At the intersection of this line with the expansion line, $H_{press} = H_{exp}$; and combining Equations (4.9) and (4.10) allows the value of S_{int} to be determined:

$$S_{int} = -\frac{a_0 - b_0}{a_1 - b_1}$$

when

$$H_{int} = b_0 + b_1 S_{int} \qquad (4.11)$$

From the HS coordinates at the intersection of the expansion line and the pressure line within the wet zone, and the set of properties a thru f at TEP, the following additional properties (or parameters) can be calculated:

g. Enthalpy at TEP (i.e., UEEP): see H_{int} and Equation (4.11)
h. Dryness fraction at intersection
i. Specific volume of vapor at intersection
j. Adjusted design exhaust flow rate based upon the change of enthalpy at TEP about its design value; see Equation (4.14)
k. Annulus velocity
l. Annulus exhaust loss using polynomial 5 or 6 from Section 4.7
m. Enthalpy at ELEP, equivalent to UEEP (parameter g) minus annulus loss (parameter l)

The *moisture fraction* (MF) at the intersection can be calculated from:

$$MF_{int} = 1.0 - \frac{H_{int} - H_{liq}}{H_{vap} - H_{liq}} \tag{4.12}$$

The *specific volume* at the intersection (SPV_{int}) can be calculated from the specific volume at the vapor boundary (parameter b), as follows:

$$SPV_{int} = SPV_{vap} \times (1.0 - MF_{int}) \tag{4.13}$$

The design value of the *exhaust flow* at the current level of power generation (EXF_{des}) can be calculated using the regression coefficients of polynomial 1 of Section 4.7. Similarly, the design value of UEEP (UP_{des}) can be obtained using the regression coefficients of polynomial 3 of Section 4.7.

An adjusted exhaust flow, to accommodate any change in turbine end point due to a deviation in back pressure about design, can be approximated by multiplying the design flow EXF_{des} by the ratio of the differences between throttle enthalpy $H_{throttle}$ and turbine end-point enthalpies:

$$EXF_{adjust} = EXF_{des} \times \frac{H_{throttle} - UP_{des}}{H_{throttle} - H_{int}} \tag{4.14}$$

From the area of the annulus (A_{ann}), the annulus velocity VEL_{ann} can be found from:

$$VEL_{ann} = \frac{EXF_{adj} \times SPV_{int}}{A_{ann} \times 3600} \tag{4.15}$$

The corresponding *annulus loss* $LOSS_{ann}$ can be calculated by using the velocity (VEL_{ann}) with the regression coefficients of polynomial 5 or 6 of Section 4.7.

Finally, the enthalpy at the back pressure (H_{bp}), or at the end point corresponding to this value of turbine end-point pressure, can be computed from:

$$H_{bp} = H_{int} - LOSS_{ann} \tag{4.16}$$

5. A regression database is constructed at the end of each of the previous four steps, consisting of:
   - Turbine end point (TEP) pressure (parameter p)
   - Temperature at TEP (parameter a)
   - Enthalpy on expansion line intersection at TEP (parameter g)
   - Enthalpy at ELEP corresponding to TEP (parameter m)
6. This database is now used to calculate the following sequence of regression coefficients:
   (i) Temperature at TEP (parameter a) as a third-order function of the

enthalpy at TEP (parameter g). This is the fundamental relationship between enthalpy and temperature along the expansion line.

The temperature on the expansion line (parameter (t)), corresponding to the enthalpy at the ELEP (parameter m), may now calculated from the set of temperature vs. enthalpy regression coefficients calculated in (i) and this temperature added to the data base

   (ii) Regression coefficients resulting from regressing enthalpy parameter g against temperature parameter (t). These third order regression coefficients will allow the turbine end-point enthalpy (UEEP) to be calculated in future from shell temperature.

   (iii) Regression coefficients resulting from regressing enthalpy parameter m against temperature parameter (t). These regression coefficients allow the expansion line end-point enthalpy (ELEP) to be calculated in future from shell temperature.

Regression coefficient sets (ii) and (iii) are those accessed by condenser performance models and also used in the estimation of cooling-water flow rate.

4.9 ESTABLISHING THE CONDITIONS WHEN THE ANNULUS BECOMES CHOKED

As stated in Section 4.5.1, the annulus at the exhaust from the low-pressure stage of a steam turbine is considered choked when the ratio of the pressure at the throat divided by the pressure at the inlet falls below 0.58. Such conditions tend to be experienced during winter months, when the cooling-water temperatures are generally low. When choking occurs, the exhaust flow to the condenser cannot rise above the annulus flow corresponding to choked conditions, while the vapor heat to be removed by the condenser is a function of the enthalpy at the throat of the annulus corresponding to these conditions. Clearly, when the exhaust flow and enthalpy become divorced from changes in the back pressure within the condenser, the condenser duty corresponding to choked conditions becomes a boundary condition to be observed during the convergence of the condenser/turbine LP stage models to be described in Chapter 5. Thus, the objective of this part of the expansion line analysis is to determine, for a given *load*, the flow through the annulus at which choking occurs and the pressure and enthalpy at the throat of the annulus under these conditions.

To determine these criteria, a database first has to be created over a narrow range of turbine end-point pressure (e.g., from 3.0 down to 0.5 in-Hg). The enthalpy and specific volume are calculated for each pressure, using Equations (4.11) and (4.13), respectively, for the conditions which exist at the intersection of the expansion line with the corresponding pressure line lying within the wet zone. Third-order polynomial regression coefficients can then be calculated for:

   (iv) Enthalpy at TEP (i.e., UEEP) versus TEP
   (v) Specific volume versus TEP

For any given turbine end-point pressure below 3.0 in-Hg, the search procedure first uses Equation (4.14) to calculate the adjusted exhaust flow EXP_{adjust} as a function of load and UEEP deviation from the design value. It then calculates the theoretical flow through the annulus ($W_{annulus}$) if the pressure at the throat is 0.58 times the inlet pressure. Under nonchoked conditions, the annulus flow $W_{annulus}$ must be greater than the adjusted exhaust flow EXF_{adjust}. Choking is defined as having occurred when the difference between these two flows falls to zero; that is,

- If $(W_{annulus} - EXF_{adjust}) > 0.0$, then flow is not choked.
- If $(W_{annulus} - EXF_{adjust}) \leq 0.0$, then flow is choked.

Lewitt [1953] shows that, for a nozzle, the throat velocity is proportional to the square root of the adiabatic (or isentropic) heat drop between inlet and throat. This enthalpy difference can be obtained from the enthalpy on the expansion line corresponding to pressures TEP and 0.58(TEP), respectively, using regression coefficients (iv); while coefficients (v) can be used to calculate the specific volume at the throat as a function of 0.58(TEP). The throat velocity (in ft/s) can be calculated from:

$$VEL_{throat} = 224.0 \sqrt{H_{TEP} - H_{throat}} \tag{4.17}$$

and annulus mass flow from:

$$W_{annulus} = \frac{3600 A_{ann} \times 224 \sqrt{H_{TEP} - H_{throat}}}{SPV_{throat}} \tag{4.18}$$

A Newton-Raphson search (see Section A.2.3 of Appendix A) is then conducted, using TEP as the independent variable, the flow difference (W_{diff}) being tested at each iteration:

$$W_{diff} = |EXT_{adjust} - W_{annulus}| < W_{thresh} \tag{4.19}$$

The solution to this equation is found when the absolute value $|W_{diff}|$ of this mass flow difference falls below an assigned threshold value W_{thresh} (e.g., 200 lbm/h). On convergence, the parameters passed to the condenser performance model are:

P_{throat}	Throat pressure when choking occurs = 0.58(TEP)	in-Hg
H_{throat}	Enthalpy at throat when choking occurs	Btu/lb_m
$W_{annulus}$	Annulus mass flow when choking occurs	lb_m/h

4.10   CONDENSER DUTY FROM CURRENT CONDITIONS

Using the results from the expansion line analysis, the condenser duty can be calculated for a given set of turbogenerator operating conditions by multiplying the exhaust flow by the vapor heat to be removed, i.e., by the difference between the end point enthalpy and enthalpy of water at the same pressure. The details depend on the manner in which the unit is being operated; there are two cases: boiler-follow mode, and turbine-follow mode.

4.10.1   Boiler-Follow Mode

All nuclear plant turbogenerators and some fossil fuel plants operate in the boiler-follow mode, for which the amount of steam generated determines the exhaust flow EXF, the relationship between these flows being obtained by regression analysis of design thermal kit data. Similarly, the design value of power can be regressed with respect to the throttle or exhaust flow, as can the design value of UEEP. These sets of regression coefficients can be identified as:

    (vi) Turbine exhaust flow versus throttle flow
    (vii) Design value of power versus exhaust flow
    (viii) Design value of UEEP versus exhaust flow

In the boiler-follow mode, the amount of power being generated at any time can be estimated, first, as a function of exhaust flow, this value then being adjusted in accordance with the deviation in UEEP about the design value.

Regression coefficients (vii) are first used to provide the design generated power MW_{des} as a function of exhaust flow. The design value of UEEP (UP_{des}) can then be calculated by presenting regression coefficients (viii) to a third-order polynomial with exhaust flow as the independent variable. The actual UEEP (UP_{act}) is the enthalpy on the expansion line determined as a function of current back pressure. It can be calculated using regression coefficients (ii) and presenting them to a third-order polynomial with the independent variable the saturation temperature corresponding to present back pressure.

An estimate of power being generated (MW_{est}) can now be obtained from:

$$MW_{est} = MW_{est} + \frac{EXF \times (UP_{des} - UP_{act})}{3,412,900} \qquad (4.20)$$

Regression coefficients (iii), developed during expansion line analysis, will allow the expansion line end-point enthalpy H_{ELEP} to be determined as a function of back pressure. The liquid enthalpy corresponding to the back pressure (H_{liq}) can be obtained from the steam tables. Finally, the condenser duty Q contributed by the exhaust heat from the turbine can be estimated from:

$$Q = EXF \times (H_{ELEP} - H_{liq}) \qquad (4.21)$$

4.10.2 Turbine-Follow Mode

Most of the turbogenerators installed in fossil fuel plants operate in the turbine-follow mode, in which the load is determined by setting the governor, the fuel firing rate to the boiler then being adjusted to maintain the pressure and temperature of the steam at the turbine throttle at their desired values. Since the amount of power to be generated is a given, the governor will adjust the throttle, and hence exhaust, flows about their design values as the back pressure fluctuates, the adjusted exhaust flow EXF_{adjust} being calculated in accordance with Equation (4.14).

As in the boiler-follow mode, regression coefficients (iii), developed during expansion line analysis, will allow the expansion line end-point enthalpy H_{ELEP} to be determined as a function of back pressure. The liquid enthalpy corresponding to the back pressure (H_{liq}) can be obtained by consulting the steam tables. Condenser duty when in the turbine follow mode can now be estimated from:

$$Q = EXF_{adjust} \times (H_{ELEP} - H_{liq}) \qquad (4.22)$$

4.10.3 Condenser Duty under Choked Flow Conditions

If the back pressure should fall below P_{throat}, then the annulus is choked. Condenser duty is now a function of the parameters calculated in Section 4.9, while H_{liq} is the enthalpy corresponding to the actual back pressure. Thus, the duty under choked conditions will be:

$$Q = W_{annulus} \times (H_{throat} - H_{liq}) \qquad (4.23)$$

4.11 OTHER CONTRIBUTORS TO CONDENSER DUTY

The total duty on a condenser may not be only the amount of latent heat to be removed from the turbine exhaust as a function of a given combination of load and back-pressure conditions. Sometimes, the exhaust from boiler feed pump turbines is also discharged into one or more condenser compartments; and there is always a relatively small amount of latent heat to be removed from various drains. While the latter is comparatively insignificant, the latent heat to be removed from boiler feed pump turbine exhausts can be quite large. Thus, in order to estimate the cooling-water flow rate (GPM) with greater accuracy, the total amount of latent heat to be removed from all sources (Q_{tot}) must be known.

4.12 COOLING-WATER FLOW RATE ESTIMATION

A confident knowledge of condenser duty can be used to estimate the amount of cooling water flowing through the condenser tubes. Table 1.1 listed sets of regression coefficients which enable the density ρ and specific heat C_p of fresh water to be calculated as a function of temperature. Appendix B contains a method of computing the same properties for seawater. If the inlet temperature T_{in} and outlet temperature T_{out} of the cooling water as it enters and leaves the condenser are both known, then a first approximation to the cooling-water flow rate, based on condenser duty, can be obtained from:

$$\text{GPM} = \frac{Q_{TOT} \times 62.3}{8.34 \times 60 \times (T_{out} - T_{in})C_p\rho} \qquad (4.24)$$

Chapter 6 contains some case studies in which this equation has been used to estimate cooling-water flow rate, and with some interesting results. Clearly, a more accurate estimate of this flow rate can always improve the quality of calculated values of heat transfer coefficients, fouling resistances, and cleanliness factors.

Where tubesheet fouling is a problem, an estimate of cooling-water flow rate can also indicate the progress of this type of fouling. It is also recognized that the cooling-water flow rate calculated from the pump characteristic curve as a function of pump discharge pressure is seldom strictly proportional to the number of pumps currently in service. However, using this flow estimation method has successfully captured the real effect on total flow when different numbers of cooling-water pumps are operating in parallel.

Chapter 5

INTERACTIVE MODEL OF CONDENSER AND LOW-PRESSURE TURBINE STAGE

So far in this book, performance monitoring of a steam turbogenerator condenser has been discussed as an engineering task, but the economic implications of the resulting data can have a significant impact on how the unit should be operated. Judgments about economic performance are usually based on the amount of heat lost to the cooling water, combined with any changes in power generation attributable to changes in condenser performance. Changes in condenser losses can be estimated by comparing the condenser's present performance with its performance in the clean design state for the same load, cooling-water flow, and inlet temperature conditions. Knowing the magnitude of the loss difference can help determine the source of changes in condenser performance and also help quantify the operating cost to the plant of these possibly avoidable losses.

Chapter 4 discussed how one set of design conditions can be abstracted from the "thermal kit" normally provided by the turbine builder for the unit. Another set of conditions can be obtained from the condenser design data sheet, but since thermal kits take no account of condenser configuration, these two sets of data have to be reconciled.

Past practice for monitoring condenser performance has usually been to treat the condenser as if it were functioning in isolation and independent of any interaction with the turbogenerator. However, thermal kits also normally include curves, similar to Figures 2.1 and 2.2, that indicate the changes in heat rate or in power generated due to variations in condenser back pressure about the design value, their magnitude depending on the exhaust flow rate and, hence, load. These figures suggest that, should conditions external to the condenser, or even within the condenser, affect back pressure, then the condenser and turbogenerator will respond by interacting with one another. Those external conditions which can affect back pressure are cooling-water flow rate, cooling-water inlet temperature, exhaust flow rate, exhaust enthalpy, and air ingress into the condenser shell, as well as the degree of fouling.

The mode in which the turbogenerator operates can also affect the cause/effect relationships between the parameter sets. For instance, as discussed in Section 4.10, with a turbogenerator unit operating in the *boiler-follow* mode, variations in condenser back pressure will not only affect the amount of heat lost to the cooling water but also cause a change in the power being generated, and this must also be taken into account in the economic evaluation. However, when a unit is operating in the *turbine-follow* mode, variations in condenser back pressure will still affect the amount of heat lost to the cooling water, but the turbine governor will now adjust the amount of steam passing through the throttle valve (and, thus, the LP turbine exhaust flow), in order to maintain the set level of generated power. An improved condenser performance monitoring program must take these interactions into account.

5.1   TURBINE/CONDENSER MODEL TYPES— BOILER-FOLLOW MODE

The turbine model for the boiler-follow mode has been outlined in Chapter 4 in terms of the expansion line analysis. The basic thermal kit data is regressed with respect to turbine exhaust flow or throttle flow, the latter being the primary independent variable. The different condenser model configurations include:

- Single-pass, single-compartment condenser
- Two-pass, single-compartment condenser
- Multiple-compartment condenser, the water flowing in series through the compartments, with each compartment operating at a different back pressure
- Several single-compartment condensers, each of the single-pass or straight-through type, arranged in series or in parallel, with each compartment possibly operating at a different back pressure

In all of the condenser models for turbogenerators operating in the boiler-follow mode, the exhaust flow from the LP turbine stage is a boundary condition, while the power assumes an equilibrium value when the model converges.

5.2   TURBINE/CONDENSER MODEL TYPES— TURBINE-FOLLOW MODE

The turbine model for the turbine-follow mode has, again, been outlined in Chapter 4 in terms of expansion line analysis. In this mode, the basic thermal kit data is regressed with respect to generator power, which is the primary independent parameter. The types of condenser model configuration for this operating mode are the same as for the boiler-follow mode. However, in all of the condenser models for turbogenerators operating in the turbine-follow mode, the power is a boundary condition; while the flow and enthalpy of the exhaust from the LP turbine stage assume an equilibrium value when the model converges.

5.3 COMPARTMENTAL RELATIONSHIPS

Since the operating condition and associated performance of a condenser change very slowly over time, modeling only the steady-state behavior will provide adequate and reliable data for judging them with confidence. Putman [1995] showed that such models can be constructed compartment by compartment. With a single-compartment condenser, the cooling-water inlet flow and temperature are givens (i.e., boundary conditions), these parameters also being included in the set of operating variables used by the model. However, with multicompartment condensers having their compartments arranged in series, while the water flow is the same for all compartments, the inlet water temperature to a later compartment is the outlet water temperature from the preceding compartment. Also when the water flow to multiple compartments is in series, each compartment has a different back pressure, so that the temperature differences across the tube walls will vary from compartment to compartment. Finally, while there are some parameters common to all compartments, other parameters are intimately associated with the conditions in a particular compartment. The set of parameters needed to model a particular condenser also depends on whether the unit is being operated in a boiler-follow or turbine-follow mode.

However, even with these variations of parameter combinations, the set of mass/heat balance equations is essentially the same for each compartment. There are three different heat transport quantities which have to be numerically identical when the compartment conditions are in equilibrium, namely:

Q_1 Latent heat of condensation in the exhaust vapor associated with a given load at equilibrium back pressure
Q_2 Amount of heat transferred through the tube walls
Q_3 Amount of heat acquired by the cooling water

Under equilibrium conditions,

$$Q_1 = Q_2 = Q_3 \tag{5.1}$$

While there is an essentially linear relationship between Q_3 and its associated set of variables, the relationships between Q_1 and Q_2 and their sets of variables are nonlinear, an important consideration when selecting the method for solving the model equations. The Newton-Raphson method has long been known as a robust and accurate method for solving sets of linear and nonlinear simultaneous equations, and Putman and Saxon [1996] showed how it has influenced the structures of the LP turbine/condenser models outlined in this chapter. The method is also described in Section A.3 of Appendix A.

5.3.1 Latent Heat of Condensation Q_1

As discussed in Chapter 4, the heat of condensation to be removed from the exhaust vapor, i.e., condenser duty, is a complicated function of back pressure and either generator load or exhaust flow, depending on the operating mode of the turbine. Equations (4.21), (4.22), and (4.23) define the corresponding method for calculating the duty.

In a single-compartment condenser, the flow to be used when calculating condenser duty is the exhaust flow specified in the turbine thermal kit data. However, with multicompartment condensers, it is customary for the turbogenerator thermal kit to specify the *total* exhaust flow to all compartments for a given load. To obtain the exhaust flow to each compartment, Silvestri [1995] suggests that, since the variation in pressure differential between the IP or LP inlet stage and compartment back pressures is relatively small, a reliable exhaust flow estimate can be obtained by dividing the total exhaust flow by the number of compartments.

Q_1 for Boiler-Follow Mode

In Section 4.10.1 it was shown that:

$$Q_1 = EXF \times (H_{ELEP} - H_{liq}) \qquad (4.21)$$

where

$$
\begin{aligned}
EXF &= \text{exhaust flow calculated using regression} \\
&\quad \text{coefficients (vi) of Chapter 4} \\
H_{ELEP} &= \text{end-point enthalpy calculated as a function of} \\
&\quad \text{back pressure using regression coefficients (ii)} \\
&\quad \text{obtained during expansion line analysis} \\
H_{liq} &= \text{liquid enthalpy corresponding to the condenser} \\
&\quad \text{back pressure, obtained by consulting the steam tables}
\end{aligned}
$$

Q_1 for Turbine-Follow Mode

In Section 4.10.2 it was shown that:

$$Q_1 = EXF_{adjust} \times (H_{ELEP} - H_{liq}) \qquad (4.22)$$

where

$$
\begin{aligned}
EXF_{adjust} &= \text{exhaust flow calculated in accordance} \\
&\quad \text{with Equation (4.14)}
\end{aligned}
$$

and H_{ELEP} and H_{liq} are as defined for the boiler-follow mode.

5.3.2 Tube Heat Transfer Rate Q_2

The basic relationship for the heat transfer through the tube walls in a cross-flow condenser follows a modified form of Equation (1.21):

$$Q_2 = \frac{PF \times UA(T_{out} - T_{in})}{\ln\left(\dfrac{T_v - T_{in}}{T_v - T_{out}}\right)} \tag{5.2}$$

To calculate the heat transfer coefficient U as a function of current condenser operating conditions requires that the ASME method be used. The procedures outlined in Sections 1.1, 1.2, and 1.4 calculate the required set of thermal resistances, these being substituted in a form of Equation (1.15) to calculate the clean-tube heat transfer coefficient:

$$U = \frac{1.0}{1/h_f + R_w + R_f} \tag{5.3}$$

The performance factor PF included in Equation (5.2) was discussed in Section 2.5. Its value at design duty can be derived from the data contained in the design data sheet for the condenser. The variation in performance factor with respect to load, as discussed in Section 2.4, can also be quantified by calculating the performance factor, i.e., the effective U coefficient divided by the ASME U coefficient, both being obtained from data taken from the unit when operated over a wide range of loads and with the condenser just cleaned. A regression analysis of these calculated performance factors with respect to load will provide the values of constants a_0 and a_1 included in the first-order linear equation

$$PF = a_0 + a_1 \times MW \tag{5.4}$$

It is these values of PF calculated from Equation (5.4) as a function of load which should be used in Equation (5.2)

It is, of course, assumed that the actual measured temperatures used to calculate the effective U coefficient are truly representative of the operating conditions and, with properly calibrated instrumentation, this is normally found in practice to be the case.

5.3.3 Heat Acquired by Cooling Water Q_3

The relationship between the heat acquired by the water (Q_3) and the water flow and water temperature rise may be obtained by rearranging Equation (4.24):

$$Q_3 = GPM \times 8.34 \times 60 C_p \frac{\rho}{62.3}(T_{out} - T_{in}) \tag{5.5}$$

The flow (GPM) can be either the design flow or else the flow calculated in accordance with Equation (4.24). The density of the water (ρ, or DENS) and its specific heat (C_p, or SPHT) can be obtained by applying the regression coefficients in Table 1.1 to a third-order polynomial in terms of the bulk temperature, i.e., the mean of the water inlet and outlet temperatures.

5.4 EQUATION MATRIX STRUCTURE

For speed of program execution, it is important to include as few variables as possible in the solution matrix. Since the turbine attributes are defined in terms of generated power in the form of expansion line analysis, a single-compartment condenser requires only five variables and five equations, the set of model variables and equations being contained in Table 5.1.

Table 5.1. Newton-Raphson Data Structure for Single-Compartment Condenser

Change vector **G**					
ΔTFLOW	ΔTIN	ΔEXH1	ΔSTMP1	ΔTOUT1	
1					-err(1)
	1				-err(2)
		1	$\partial f(3)/\partial STMP1$		-err(3)
$\partial f(4)/\partial TFLOW$	$\partial f(4)/\partial TIN$	$\partial f(4)/\partial EXH1$		$\partial f(4)/\partial TOUT1$	-err(4)
$\partial f(5)/\partial TFLOW$	$\partial f(5)/\partial TIN$	∂ ∂	$\partial f(5)/\partial STMP1$	$\partial f(5)/\partial TOUT1$	-err(5)
Matrix of partial differentials					Error vector **E**

Set of Model Variables

Model variables fall into two categories: variables common to all compartments, and those assigned to a specific compartment. In the case of a single-compartment condenser, the set of variables consists of the following:

Common Variable

1. Cooling-water flow rate   TFLOW   GPM

Unique to First Compartment

2. Cooling-water inlet temperature	TIN	°F
3. Exhaust latent heat flow	EXH1	MBtu/h
4. Shell vapor temperature	STMP1	°F
5. Cooling-water outlet temperature	TOUT1	°F

Note that the operating variables included in the Newton-Raphson procedure must all be initialized at the beginning of the program with a consistent set of data, especially temperature data. Because log mean temperature differences are calculated within the model, using Equation (5.2), it is especially important that the water inlet temperature be lower than the outlet temperature, and that the shell vapor temperature be higher than both. An appropriate set of initial values is usually those presently being measured on the unit.

Set of Model Equations

The five model difference equations reflected in the matrix contained in Table 5.1 are as follows:

Common to All Compartments

1. TFLOW – assigned water flow $\qquad = -0$

Unique to First Compartment

2. TIN – assigned inlet temperature $\qquad = -0$

3. Q_1 – EXH1 $\qquad = -0$

4. Q_1 – Q_3 $\qquad = -0$

5. Q_2 – Q_3 $\qquad = -0$

Note that, while the right-hand sides of these difference equations should be zero at convergence, there will in fact be residual values calculated at each iteration. Note also that, conceptually, it is the negative of each calculated difference which is assigned to the corresponding element of the error vector of Table 5.1.

5.4.1 Water Flow Equation

The assigned cooling-water flow to the condenser ($FLOW_{ref}$) will be either the design value for the condenser, allowing for the number of pumps in operation, or the value calculated in accordance with Equation (4.24). The variable TFLOW, used in the Newton-Raphson matrix, is initially assigned the value of $FLOW_{ref}$, and during each iteration of the Newton-Raphson procedure, the following calculation is performed:

$$\text{TFLOW} - \text{FLOW}_{\text{ref}} = -\text{error}(1)$$

with TFLOW later adjusted at the end of each iteration in accordance with:

$$\text{TFLOW} = \text{TFLOW} + \Delta\text{TFLOW}.$$

The partial differential of TFLOW in this difference equation is unity, and it is seen that a 1 is placed in the TFLOW column of the matrix in Table 5.1 corresponding to the row assigned to difference equation 1. At convergence, it will be found that the value of TFLOW is very close to the assigned flow value FLOWref and ?TFLOW is within its assigned tolerance.

5.4.2 Cooling-Water Inlet Temperature Equation

The assigned cooling-water condenser inlet temperature will be the value currently being measured (T_{ref}). The variable TIN, used by the Newton-Raphson matrix, is initially assigned the value T_{ref}, and during each iteration of the Newton-Raphson procedure, the following calculation is performed:

$$\text{TIN} - T_{\text{ref}} = -\text{error}(2)$$

but this result is later adjusted at the end of each iteration in accordance with

$$\text{TIN} = \text{TIN} + \Delta\text{TIN}$$

The partial differential of TIN in this difference equation is unity, and, as shown in Table 5.1, a 1 is placed in the TIN column of the matrix corresponding to the row assigned to difference equation 2. At convergence it will be found that the value of TIN is very close to the assigned value of water inlet temperature, T_{ref}.

5.4.3 Exhaust Latent Heat Flow

When a unit is operating in the turbine-follow mode, the heat of condensation to be removed from the LP exhaust (Q_1) is a function of load and condenser back pressure, as reflected in Equation (4.22). The exhaust flow rate can be computed using Equation (4.14), while the enthalpy of the vapor at condenser back pressure (H_{ELEP}) can be calculated using regression coefficients (iii) of Section 4.8 and the liquid enthalpy H_{liq} corresponding to the back pressure can be obtained from the steam tables.

Initially, EXH1 is made equal to Q_1. Then, during each iteration of the Newton-Raphson procedure, EXH1 is compared with the value of Q_1 calculated from the present values of the variables, the difference being stored in the error vector:

$$\text{EXH1} - Q_1 = -\text{error(3)} \tag{5.6}$$

At the end of each iteration of the Newton-Raphson procedure, the value of STMP1 is updated in accordance with:

$$\text{STMP1} = \text{STMP1} + \Delta\text{STMP1}$$

while EXH1 is also updated in accordance with:

$$\text{EXH1} = \text{EXH1} + \Delta\text{EXH1}$$

At the beginning of the next iteration, this new value of STMP1 and the corresponding back pressure are used to recalculate the exhaust latent heat flow Q_1, and it is this new value which is compared with the updated value EXH1 during this iteration. Now:

$$Q1 = \text{EXF}_{\text{adjust}}(\text{H}_{\text{ELEP}} - \text{H}_{\text{liq}}) = f(\text{MW}, \text{STMP1}) \tag{5.7}$$

To differentiate Equation (5.6), a 1 must be placed in the ΔEXH1 column of the matrix, and the negative of the partial differential $(\partial Q1/\partial\text{STMP1})$ must be calculated for a small perturbation about the present value of STMP1, using Equation (4.22). This value of the partial differential must now be stored in the ΔSTMP1 column of the matrix corresponding to equation 3.

At convergence, the calculated value of ΔEXH1 will lie within the tolerance assigned to variable EXH1, while STMP1 will assume an equilibrium value.

5.4.4 Equality between PQ_1 and Q_3

For a given load, the value of Q_1 calculated in accordance with Equation (4.22) can be considered to be a function of load and shell temperature STMP1, as shown in Equation (5.7). Subtracting Q_3 from Q_1, the difference equation becomes:

```
Q1 - Q3 = f(MW,STMP1)
        - (GPM * 8.34 * 60 * SPHT * (Tout - Tin) * DENS/62.3)
        = -error(4)
```

$$\tag{5.8}$$

The partial differential $\partial Q1/\text{STMP1}$ must now be placed in the ΔSTMP1 column of the matrix corresponding to equation 4, while the partial differentials $\partial Q3/\partial T_{\text{out}}$ carrying the negative sign will be placed in the ΔT_{out} column of the matrix; but the partial differential $\partial Q3/\partial T_{\text{in}}$ carrying a positive sign will be stored in the ΔT_{in} column of the matrix corresponding to equation 4. The partial differential of the expression with respect to GPM (TFLOW), i.e. $\partial Q3/\partial \text{TFLOW}$, must also be calculated and inserted in the ΔTFLOW column of the matrix corresponding to equation 4.

5.4.5 Equality between Q_2 and Q_3

The basic difference equation for this equality is:

$$Q2 - Q3 = -\text{error}(5) \tag{5.9}$$

Using Equations (5.2) and (5.5), this can be expanded into:

```
Q2 - Q3 = (PF * U * A * LMTD)
        - (GPM * 8.34 * 60 * SPHT * (T_out - T_in) * DENS/62.3)
        = -error(5)
```
(5.10)

Note that both U and LMTD are functions of all three temperatures STMP1, TIN, and TOUT, while U also varies as a nonlinear function of GPM (TFLOW) in the case of Q_2 but a linear function of flow in the case of Q_3. The partial differentials of error(5) with respect to these four variables are best obtained by perturbing each variable in turn by a small amount and computing the resulting change in Equation (5.9), divided by the magnitude of the perturbation. These calculated partial differentials must then be summed with respect to each variable, the sums being stored in the appropriate column/row element of the matrix.

5.5 PRACTICAL CONSIDERATIONS

- When modeling multicompartment condensers, it is helpful to consider the sparse matrix shown in Table 5.1 as the first module in a set of linked modules. In a complex model, the first module should be defined and verified. Once this has been completed, the matrix should be expanded by adding the variables and equations for the second model and then verifying the behavior of the expanded matrix. Additional modules should be added one by one until the matrix is complete for the condenser configuration being analyzed.
- If the execution of the model should come to a halt, it is usually because (1) the equations have not been correctly defined, (2) the set of boundary conditions is inadequate, or (3) the differentiation of an equation has not been performed correctly.
- While the Newton-Raphson method has been found to be robust and accurate, it is not perfect! A convergence tolerance is assigned to each individual variable, and to achieve convergence, the absolute values of all changes in variables must each lie below the assigned threshold tolerance. Unfortunately, if these tolerances are made too narrow, not only may convergence not take place, but divergence may even occur. With this in mind, satisfactory tolerances have been found to be $\pm 0.2°F$ on temperatures; $\pm 1000 \text{lb}_m/\text{h}$ on flow; ± 1000 Btu/h on duty; and $10^{-6}°F/[(\text{Btu}/(\text{ft}^2 \cdot \text{h})]$ on fouling resistance.
- At the end of each iteration, after the variables have been updated by the calculated changes, each of their values should be monitored, especially

to ensure that all temperature differences are positive. If this check is not performed and attempts by the program to calculate the logarithm of a negative number create an error condition, the program can come to a halt and so will not converge.

5.6 AFTER CONVERGENCE

Once the model has converged, a complete set of variables, together with the values of the ASME heat transfer coefficients, will become available. From this information, the data to be contained in the condenser performance report can be abstracted and/or calculated. The magnitude of the possibly avoidable heat losses to the condenser cooling water as a result of fouling is one important parameter, while, in the boiler-follow mode, the change in power resulting from the change in back pressure can also be calculated, using Equation (4.20).

5.7 CALCULATION OF FOULING RESISTANCE

It is clear that, for tubes in a fouled state, Equation (5.3) can be replaced by Equation (1.15). Clearly, for a clean condenser, the fouling resistance term R_{foul} would be set to zero. To calculate the fouling resistance of any deposits in a single-compartment condenser, using current operating data, requires that the matrix of Table 5.1 be expanded. A new variable for fouling resistance (FOUL) must be added, as well as an additional equation which, at convergence, will force the shell vapor temperature (STMP1) to equal the temperature corresponding to the present back pressure (STMP1$_{ref}$). The additional equation takes the form:

$$STMP1 - STMP1_{ref} = -err(6)$$

Since the analysis of the LP stage expansion line allows the condenser duty to be calculated as a function of load and back pressure, the duty at convergence will be equal to the duty presently being measured for the fouled condenser.

Further, since the cooling-water flow rate is calculated from the duty and water temperature rise, the water outlet temperature should also be close to its present value at convergence, and the fouling resistance calculated will be that due only to deposits on the tube inside walls. Thus, Table 5.1 would now be expanded to the form of Table 5.2.

5.8 LOSSES DUE TO TUBESHEET FOULING

Let Equation (1.15) still be used to calculate the heat transfer coefficient in accordance with the ASME procedure, let the fouling resistance be assigned the value calculated using Table 5.2, and let the water flow rate be assigned the value it assumes when the condenser is clean. The model shown in Table 5.1 can then be used to estimate the condenser cooling-water heat losses at the increased flow

Table 5.2. Newton-Raphson Matrix Data Structure for Single-Compartment Condenser in the Fouled State

ΔTFLOW	ΔTIN	ΔEXH1	ΔSTMP1	ΔTOUT1	ΔFOUL	
1						-err(1)
	1					-err(2)
		1	∂f(3)/∂STMP1			-err(3)
∂f(4)/∂TFLOW	∂f(4)/∂TIN	∂f(4)/∂EXH1		∂f(4)/∂TOUT1		-err(4)
∂f(5)/∂TFLOW	∂f(5)/∂TIN		∂f(5)/∂STMP1	∂f(5)/∂TOUT1	∂f(5)/∂FOUL	-err(5)
			1			-err(6)

rate but with the tubes fouled. Subtracting these estimated cooling-water losses from the losses calculated with the model using Table 5.2 provides an estimate of the losses due to tubesheet fouling alone.

5.9 TWO-COMPARTMENT CONDENSER

For this discussion, the two-compartment condenser configuration of Figure 3.4 will be used. The water flow rate is the same for both compartments, but while Table 5.1 contains the variables for the first compartment, three additional variables are required for the second compartment, namely, STMP2, EXH2, and TOUT2.

Since the number of equations must be the same as the number of variables, three additional equations are also required:

$$Q1_2 - EXH2 = -\text{error}(6)$$
$$Q1_2 - Q3_2 = -\text{error}(7)$$
$$Q2_2 - Q3_2 = -\text{error}(8)$$

The procedures of Sections 5.4.3, 5.4.4, and 5.4.5 are applied, the variables being those assigned to the second compartment. As already noted, TFLOW is common to both compartments, while TIN2 is identical to TOUT1. Table 5.3 shows the configuration of the matrix for a two-compartment condenser, with an eight-element error vector (not shown) also assumed to be present.

Table 5.3. Newton-Raphson Matrix Data Structure for Two-Compartment Condenser

ΔTFLOW	ΔTIN	ΔEXH1	ΔSTMP1	ΔTOUT1	ΔEXH2	ΔSTMP2	ΔTOUT2
1							
	1						
		1	$\partial f(3)/\partial STMP1$				
$\partial f(4)/\partial TFLOW$	$\partial f(4)/\partial TIN$	$\partial f(4)/\partial EXH1$		$\partial f(4)/\partial TOUT1$			
$\partial f(5)/\partial TFLOW$	$\partial f(5)/\partial TIN$		$\partial f(5)/\partial STMP1$	$\partial f(5)/\partial TOUT1$			
					1	$\partial f(6)/\partial STMP2$	
$\partial f(7)/\partial TFLOW$				$\partial f(7)/\partial TOUT1$	$\partial f(7)/\partial EXH2$		$\partial f(7)/\partial TOUT2$
$\partial f(8)/\partial TFLOW$				$\partial f(8)/\partial TOUT1$		$\partial f(8)/\partial STMP2$	$\partial f(8)/\partial TOUT2$

5.10 THREE-COMPARTMENT CONDENSER

Following the previous discussion of the principles for modeling a two-compartment condenser, Putman [1997b] showed how the addition of a third set of variables and equations and the associated expansion of the matrix allows a model for a three-compartment condenser to be constructed.

Chapter 6

CASE STUDIES

These case studies show the practical application of the modeling concepts outlined in Chapters 4 and 5. Section 6.1 demonstrates the use of the low pressure turbine model in estimating the cooling water flow rate and the quality of those estimates. Section 6.2 illustrates the application of the interactive model of condenser and L.P. turbine stage to a tidewater plant and the richness of the performance and operating information that results.

These modeling techniques are of interest in that, in the past, engineers had only a modest amount of information for quantifying condenser performance or identifying the underlying causes of current condenser behavior. The shortage of such information limited knowledge of a unit's operating cost or power-generating capability. However, the modeling and analytical techniques described in earlier chapters allow previously hard-to-obtain data to be derived from fundamental measurements. This enriched quantity of data has helped:

- To improve the diagnosis of the underlying causes for changes in condenser performance
- To estimate the current cooling-water flow rate
- To establish the rate at which fouling progresses in a particular unit
- To calculate the cost of losses due to fouling
- To determine the effect of tubesheet fouling on reductions in cooling-water flow rate
- In conjunction with historical data, to establish an optimum condenser cleaning schedule

This chapter will describe two real-world case studies of condenser performance. The first, in Section 6.1, examines ways of calculating cooling-water flow rate. The second, in Section 6.2, is an analysis of issues related to fouling in a power plant located on a tidal waterway.

6.1 ESTIMATING COOLING-WATER FLOW RATE

Too often, the calculation of condenser cleanliness and fouling factors is based on an uncertain knowledge of cooling-water flow rate. Because the flow rate affects the value of the water velocity used in these calculations, the final results are subject to some doubt, together with any quantities—e.g., cost of avoidable losses—whose calculation also depends on this value. The method of estimating the cooling-water flow rate based on expansion line analysis, as outlined in Chapter 4, provides a more consistent basis for this fundamental performance monitoring parameter, allowing condenser performance to be quantified with greater confidence.

The procedure to be used in the cases presented in this chapter and already outlined in Chapter 4 takes the following into account:

1. The data obtained from the analysis of the vapor properties along the expansion line for the low-pressure stage of the turbine, based on the thermal kit data, is used to calculate condenser duty as a function of condenser back pressure and generated load.
2. The cooling-water flow rate can then be calculated from condenser duty and water temperature rise.
3. The calculation of the reference heat transfer coefficient U_{ref}, whether according to the HEI or the ASME procedure, can be based on the velocity corresponding to the cooling- water flow rate and the cross-sectional flow area of the tubes.
4. The effective heat transfer coefficient U_{eff} can be calculated from cooling-water flow rate and log mean temperature difference.
5. The cleanliness factor is, of course, the ratio U_{eff}/U_{ref} expressed as a percentage.

Most of the condensers involved in these cases had single-compartments with either one or two passes, generally as shown in Figures 3.1 or 3.2. The data base required to perform the analysis for condensers of these types is defined in Table 6.1.

The condenser design data set can usually be abstracted from the design data sheet provided by the condenser manufacturer. The clean cooling water flow rate is the design rate; as distinct from the cooling water flow rate that will be calculated using the L.P. stage expansion line model.

It should be noted that the HEI material factor used for the tube material should be that current at the time the condenser was designed. This was the value used in the calculation of the stated design cleanliness factor. The HEI material factors contained in the latest edition of their standard are usually higher than those published previously so that, if used, the calculated cleanliness factors will be lower than expected.

The design distribution factor is based on the ASME single tube value to be calculated from the condenser design data. Because the HEI expected heat transfer values relate to tube bundles, design distribution factors are at least 10% lower than the corresponding design cleanliness factor (See Section 2.6) The design distribution factor or performance factor (PF) is obtained by substituting the appropriate values of U_{eff} and U_{ASME} in Equation (2.25).

Some additional notes on the design data: The seawater concentration will be determined from the salinity of the cooling-water source as a percentage of the salinity of seawater. The HEI material correction factor will be found in the HEI *Standards for Steam Surface Condensers* [1995] and the thermal conductivity from the TEMA standards [1988]. The residual resistance reflects the oxide resistance or the resistance of any deposits remaining after tubes have been cleaned.

Table 6.1. Typical Modelling Data Set—Single-Compartment Condenser

Condenser design data set	
Tube Material	S S304/316
Tube outside diameter	1.0 Ins
Tube wall thickness	0.028 Ins
Tube length	27.83 Feet
Number of tubes	5492.0
Number of passes	2.0
Shell temperature	103.18 °F
C.W. Inlet Temperature	78.0 °F
C.W. Outlet Temperature	93.25 °F
Cooling water flow—Clean	41700.0 GPM
HEI Material factor	0.87
Thermal conductivity	8.7 B TU/sq.ft.h. °F
Oxide film resistance	.0 °F/(BTU/sq.ft.h)
Exhaust flow	326495.0 lb /h
Sea water concentration	.0%
Condenser duty	341.93 MBTU/h
Des. Cleanliness factor	85.0%
Unit system	1.0
Des. Distribution factor	63.50%

Turbine design data set	
Design load	72.0 MW
Design back pressure	2.0 ins.Hg
Throttle enthalpy	1490.0 BTU/lb
Maximum Generated Load	72.0 MW
Minimum Generated Load	15.0 MW
Auxiliary duty	0.0 MBTU/h
Reheat temperature	1000.0 °F
Annulus area	25.0 sq.ft
Minimum Exhaust Loss Velocity fps	325.0 fps

Turbine regression coefficients
Exhaust flow vs. MW load
Exhaust enthalpy vs. MW load
Turbine end-point enthalpy vs. MW load
Annulus loss vs. annulus velocity—low-velocity range
Annulus loss vs. annulus velocity—high-velocity range
Reheater outlet pressure vs. MW load
Intermediate-pressure stage inlet temperature
Boiler feed pump turbine exhaust heat vs. MW load

Condenser operating data
Generated power
Back pressure
Cooling-water inlet temperature
Cooling-water outlet temperature

The set of turbine data, for the parameters in the second part of the table, will be contained in the thermal kit supplied with the steam turbine. The regression coefficients (the third part of the table) are obtained by abstracting the appropriate data as a function of generated power from the heat/materials balances included in the thermal kit and regressing these data sets to obtain the minimum least-squares best fit for a third-order polynomial. In addition to all of these, a suitable set of steam-table algorithms will be required, together with sets of regression coefficients for the density, specific heat, conductivity, and viscosity of water.

Operating data (the fourth part of the table) can be acquired either from snapshots of the set of averaged operating data taken at a particular point in time with the unit operating under equilibrium conditions, or from sets of data averaged over five minutes and acquired automatically by a data acquisition system. In the latter case, preprocessing programs can be written to filter out those data sets acquired while the unit was in transition, as well as those taken when the unit was operating outside its normal range (e.g., shut down).

Figures 6.1 through 6.6 show plots of cooling water flow rate vs. load or time for a variety of turbogenerators of different sizes and located in various parts of the country. All show a consistent value of the cooling water flow rate even though generated power and/or cooling water inlet temperature vary over the period under study.

Figure 6.1 shows the mean cooling-water flow rate calculated from snapshot data (sets of data taken at several specific points in time) acquired during single-pump operation of a 185-MW unit located on the Ohio River, and Figure 6.2 shows the corresponding results when two pumps were in operation The data snapshots included measurements of cooling-water flow rate, inlet temperature, and load in megawatts, and the data has been plotted with respect to megawatt load. Figure 6.1 shows that with one pump operating, the mean flow was calculated to be 41,715 GPM, while with two pumps, it was 83,936 GPM as shown in Figure 6.2. Thus, flow was directly proportional to the number of pumps operating, but this is not always the case when pumps operate in parallel. It should also be noted that the calculated flow with two pumps was well below the total design flow of 93,000 GPM. This was an old unit and it was suspected that impeller wear was responsible for the difference.

In Figures 6.1 and 6.2 it is noteworthy that the observed cooling-water flow rate stayed essentially constant, close to the calculated mean, for the number of pumps operating, regardless of variations in the load or inlet water temperature. In other instances, variation in cooling-water flow rate does occur. Figures 6.3 and 6.4 show unfiltered data from an 85-MW unit in Kentucky plotted on a time basis. In Figure 6.3, representing the condenser before cleaning, the flow stays close to the mean of 41,000 GPM, but there are clearly deviations when the unit is passing through load changes. In Figure 6.4, the flow data calculated for the first 20 hours has a mean flow (with some variation) for two pumps of about 44,000 GPM with a design flow of 55,000 GPM. But from that point onward, and with only one pump running, the calculated flow is essentially constant at about 28,000 GPM, or half the design flow. Clearly, some filtering would be necessary to provide more meaningful flow data, and these results do bring into question the observation that, when these pumps are running in parallel, their combined flow is twice the flow from a single pump. Information on one-and two-pump operation is usually contained in the pump characteristic curves.

Figure 6.5 is a plot of data taken from a 300-MW unit in South Carolina. It shows a mean flow of 173,160 GPM on which other calculations can be based. This again was unfiltered data, and the obvious noise is probably due to the frequent load transitions. Figure 6.6 is a plot of data taken from a 500-MW unit

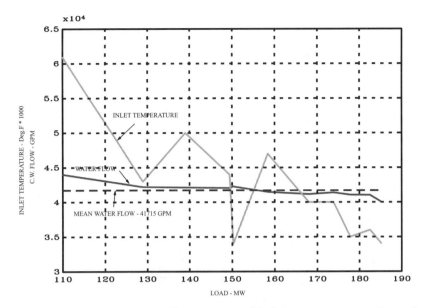

FIGURE 6.1. Cooling-water flow rate and inlet temperature plotted in relation to megawatt load for a 185-MW unit on the Ohio River with single-pump operation.

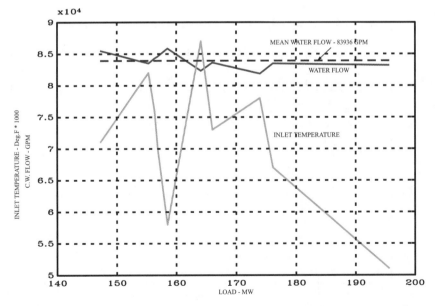

FIGURE 6.2. Corresponding plots for the unit of Figure 6.1 for two-pump operation.

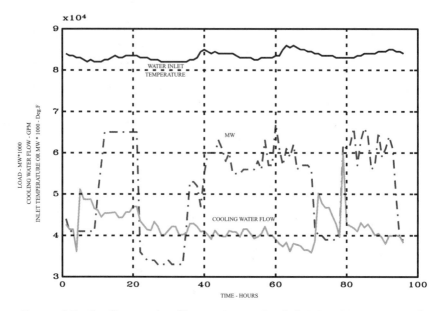

FIGURE 6.3. Cooling-water flow rate, water inlet temperature, and megawatt load plotted over time for a 70-MW unit in Kentucky before cleaning.

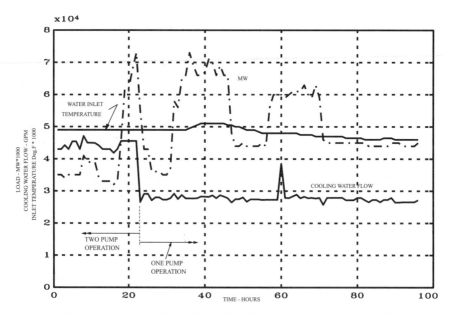

FIGURE 6.4. Corresponding plots for the unit of Figure 6.3 for the period immediately after cleaning.

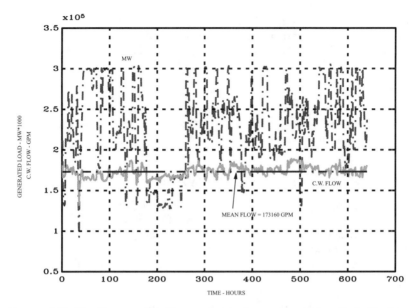

FIGURE 6.5. Cooling-water flow rate and megawatt load plotted vs. time for a 300-MW unit in South Carolina.

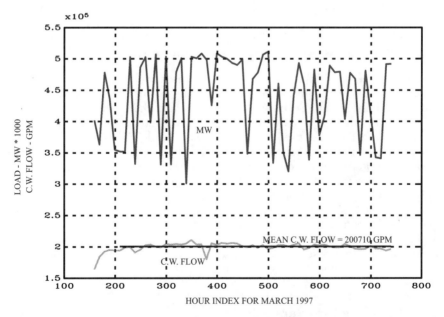

FIGURE 6.6. Cooling-water flow rate and megawatt load plotted vs. time for a 500-MW unit in Tennessee.

located in Tennessee. The mean flow of 200,170 GPM is very close to the design value of 200,000 GPM and appears to have very little noise.

These examples, together with others to be described later, indicate that the method for estimating cooling-water flow rate based upon expansion line analysis can provide interesting insights into the unmeasured changes which are occurring in the water path into and through the condenser. These changes can be due to tubesheet or deposit fouling, disparities between the design and actual water flow rates, and the behavior of multiple pumps operating in parallel compared with the same pumps when operating alone.

6.2 POWER PLANT ON TIDAL WATERWAY

The following analysis by Putman and Hornick [1998] was based on data taken from one of four coal-fired 450-MW units, all of which draw their cooling water from a saltwater bay located in Florida. The condenser has two once-through compartments (identified as 'A' and 'B') with their cooling-water paths arranged in parallel. These condensers have shown a tendency for both tubesheet fouling and deposit fouling on the inner surfaces of the tubes. The cleaning frequency varies but is on the order of days rather than weeks or months.

The data base for this unit and condenser is larger than that for single-compartment condensers and is shown in Table 6.2. Note the separate inclusion of cooling water inlet and outlet temperatures as well as shell temperature for both compartment 'A' and compartment 'B'.

Refer also to Section 6.1 for other comments concerning the set of data.

6.2.1 Estimation of Cooling-Water Flow Rate

Figure 6.7 is a plot of generated power, condenser back pressure, and cooling-water flow rate versus time, the flow rate being estimated in accordance with the method outlined in Section 4.12. In this figure, there is a clear correspondence between the rise in back pressure and a drop in the cooling-water flow rate occurring over the same time period. This downward drift in water flow rate is due mainly to tubesheet fouling, which occurs fairly rapidly: mollusks and debris have to be removed by hand during periodic unit outages scheduled for this purpose. The tubes themselves are not cleaned as often as the tubesheet, and plastic plugs are the preferred cleaning tools.

Even if the cost were affordable, the direct measurement of water flow rate would be difficult to implement because of the layout of the large-bore piping. Monitoring the cooling-water flow rate, using a computational method which can be shared among several units, is an inexpensive substitute for providing a separate instrument system for each unit.

6.2.2 Fouling Resistance Plot

The interactive condenser/turbine LP stage model outlined in Chapter 5 is used not only to estimate the cost of losses due to fouling, but also the total fouling

Table 6.2. Condenser design data set

Tube material	Allegheny Ludlum AL-6X
Tube outside diameter	0.875 in
Tube wall thickness	0.028 in
Tube length	42.25 ft
Number of tubes	17,824
Number of passes	1
Shell temperature A (i.e., back pressure)	123.95 °F
Shell temperature B (i.e., back pressure)	124.18 °F
CW inlet temperature A	90.57 °F
CW inlet temperature B	90.57 °F
CW outlet temperature A	107.05 °F
CW outlet temperature B	107.05 °F
Cooling-water flow—clean	249,200 GPM
Low-pressure stage exhaust flow	2,062,444 lb_m/h
Seawater concentration	100.0%
Condenser duty	1745.43 MBtu/h
Design cleanliness factor	85.0%
HEI material correction factor	0.850
Thermal conductivity*	7.9 Btu/(ft · h · °F)
Design distribution factor	74.96%
Residual resistance	10^{-6} °F/[(Btu/(ft^2 · h)]

Turbine design data set	
Design load	398.9 MW
Design condenser back pressure	1.73 in-Hg
Throttle enthalpy	1480 Btu/lb_m
Maximum generated load	478 MW
Minimum generated load	112.6 MW
Cooling-water flow—fouled	180,000 GPM
Auxiliary duty	0 MBtu/h
Throttle pressure	2400 psia
Throttle temperature	1000 °F
Reheat temperature	1000 °F
Annulus area	164.4 ft^2
Minimum exhaust loss velocity	550 ft/s

Turbine regression coefficients
Exhaust flow vs. MW load
Exhaust enthalpy vs. MW load
Turbine end-point enthalpy vs. MW load
Annulus loss vs. annulus velocity—low-velocity range
Annulus loss vs. annulus velocity—high-velocity range
Reheater outlet pressure vs. MW load
Intermediate-pressure stage inlet temperature
Boiler feed pump turbine exhaust heat vs. MW load

Condenser operating data
Generated power
Back pressure—compartment A
Back pressure—compartment B
Cooling-water inlet temperature—compartment A
Cooling-water inlet temperature—compartment B
Cooling-water outlet temperature—compartment A
Cooling-water outlet temperature—compartment B

*This value for the thermal conductivity of AL-6X was that stated by the HEI standard at the time the condenser was designed. A series of experiments conducted at Purdue University found the actual value to be 6.8 Btu/(ft · h · °F), and Allegheny Ludlum has adopted that figure in its literature. However, this difference in thermal conductivity would not have a significant effect on the results.

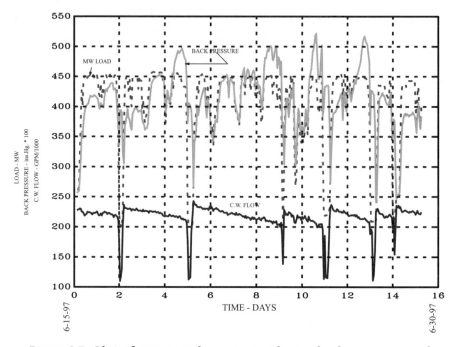

FIGURE 6.7. Plot of generated power, condenser back pressure, and cooling-water flow rate vs. time for the tidewater power plant case study.

resistance, together with the individual resistances due to both tubesheet and tube deposit fouling. It also yields the information shown in Figure 6.8 for this case study: the variation of back pressure, flow, and fouling resistance versus time, as well as the corresponding load on the generator. Regardless of the load profile over the time period, the combined fouling resistance is seen to increase at the same time as the cooling-water flow rate is seen to fall. The back pressure is also progressively rising, because the decrease in flow rate is making it more difficult to transfer the necessary condensation heat.

The corresponding fouling losses in MBtu/h are shown in Figure 6.9. These, of course, fluctuate with load, but for a given load, the losses are seen to be rising over time. If the dollar cost per MBtu of heat is known, this increase in losses can be assigned as the avoidable operational cost attributed to fouling.

6.2.3 Condenser Performance Monitoring Program

The calculation results from a program written for this tidewater plant are included in Tables 6.3 through 6.8. After initializing execution, a menu is displayed as shown in Table 6.3, which includes the following functions:

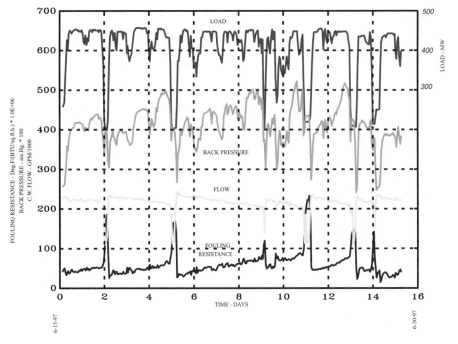

FIGURE 6.8. Variation of back pressure, cooling-water flow rate, and fouling resistance vs. time for the tidewater power plant case study.

F1 Allows the design data file for the unit under study to be selected.

F2 Brings up the display of the operating data set, as shown in Table 6.4. The upper field of this display shows the set of significant design data for this unit, while the lower field is the area through which the set of current operating data can be loaded.

F3 On selecting function F3 from the menu, the set of design and operating data currently loaded into the database is presented to the program, which proceeds to execute the model and display the results.

Clearly, by switching between functions F2 and F3, several sets of data can be processed one by one during a single session.

F4 Selecting this function allows the set of data for this calculation to be stored in a historical data file, which can be used in the future for analysis and plotting.

F5 A set of steam-table algorithms can be accessed by calling this function. This can be useful to verify saturation temperatures as a function of condenser back pressure, or the function can be used as an engineering utility.

F6 Calling this function terminates the program.

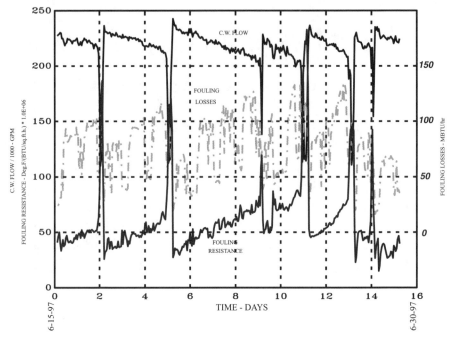

FIGURE 6.9. Fouling losses, back pressure, and cooling-water flow rate plotted vs. time for the tidewater power plant case study.

Table 6.3. Main Menu

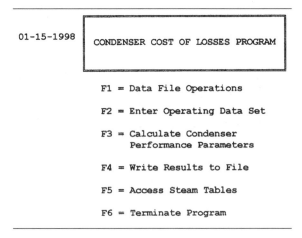

01-15-1998   CONDENSER COST OF LOSSES PROGRAM

F1 = Data File Operations

F2 = Enter Operating Data Set

F3 = Calculate Condenser
     Performance Parameters

F4 = Write Results to File

F5 = Access Steam Tables

F6 = Terminate Program

Table 6.4. Input Data Table

01-15-1998 14:23	CONDENSER COST OF LOSSES PROGRAM Input Data Set

Metal: Allegheny Ludlum AL-6X

FIXED DATA

Tube Dia.	.875	in	Sea Water Conc.	100.000	%
Wall Thick.	.028	in	Des. Duty	1745.430	MBtu/h
Tube Length	42.250	ft	Des. CW Flow	249200.000	GPM
No. Tubes	17824.000		Des. Dis. Fact	74.964	%
No. Passes	1.000		Shell Temp. A	123.950	Deg.F
HEI Matl. Factor	.850		Shell Temp. B	124.180	Deg.F
Thermal Cond.	7.900	Btu/etc.			

VARIABLE DATA

Load	398.941	MW	CW Temp. Out	107.050	Deg.F
CW Temp. In, A	90.570	Deg.F	Back Press. A	3.844	in-Hg
CW Temp. In, B	90.570	Deg.F	Back Press. B	3.868	in-Hg

Table 6.5. Cooling-Water Flow Estimate Results

01-15-1998 14:23	COOLING WATER FLOW ESTIMATION PROGRAM

MW Load	=	398.94 MW
Back Pressure, A	=	3.84 in-Hg
Back Pressure, B	=	3.87 in-Hg
Cooling Water Temperature: In	=	90.57 Deg.F
Cooling Water Temperature: Out	=	107.05 Deg.F
Condenser Duty	=	1838.38 MBtu/h
Estimated Cooling Water Flow	=	228258.70 GPM

The complete set of results is displayed as several pages having the appearance shown in Tables 6.5 through 6.7. Table 6.5 lists the significant operating data available to the program for evaluation, together with the condenser duty estimated from the thermal kit data and the corresponding water flow rate calculated in accordance with Equation (4.24).

The lower left-hand column of Table 6.6 lists the present set of operating con-

Table 6.6. Condenser Performance Calculations

01-15-1998 14:23	CONDENSER COST OF LOSSES PROGRAM

Generator Load	=	398.941	MW
Fouling Losses	=	34.968	MBtu/h

		PRESENT CONDITIONS	CLEAN CONDITIONS	
CW Flow	=	228258.700	249200.000	GPM
Condenser Duty	=	1838.382	1803.414	MBtu/h
Exhaust Flow	=	1810.699	1782.084	klb/h
Inlet Temperature	=	90.570	90.570	Deg.F
Outlet Temperature, A	=	107.022	105.358	Deg.F
Steam Temperature, A	=	123.950	115.946	Deg.F
Back Pressure, A	=	3.844	3.077	in-Hg
Outlet Temperature, B	=	107.031	105.357	Deg.F
Steam Temperature, B	=	124.180	115.943	Deg.F
Back Pressure, B	=	3.868	3.077	in-Hg

Table 6.7. Heat Transfer Coefficients and Fouling Resistances

01-15-1998 14:23	CONDENSER HEAT TRANSFER COEFFICIENTS

		PRESENT CONDITIONS	CLEAN CONDITIONS
U Coefficient, A (ASME)	=	586.891	824.316
U Coefficient, A (HEI)	=	671.345	703.063
U Coefficient, A (Actual)	=	439.687	617.811
U Coefficient, B (ASME)	=	581.645	824.374
U Coefficient, B (HEI)	=	671.345	703.063
U Coefficient, B (Actual)	=	435.743	618.255
Fouling Resistance, A	=	.0004289	.0000000
Fouling Resistance, B	=	.0004434	.0000000

ditions together with the condenser duty calculated for the condenser in the fouled state. The values for cooling-water flow and condenser duty are identical to those shown in Table 6.5. The right-hand column of Table 6.6 shows the

Table 6.8.   Distribution of Fouling Losses

01-15-1998 14:23	CONDENSER FOULING LOSS DISTRIBUTION		
	Generator Load	=	398.941 MW
Total Fouling Losses	=	34.968 MBtu/h	100.00 %
Macro Fouling Losses	=	6.258 MBtu/h	17.90 %
Micro Fouling Losses	=	28.710 MBtu/h	82.10 %

expected values of the equivalent set of parameters if the unit were generating the same megawatt load with the condenser in the clean state. The water flow rate has been increased to the value it is known to assume immediately after the condenser has been cleaned. With a clean condenser, the back pressures in both compartments fall, allowing the value of the exhaust flow to fall also; while the water outlet temperatures also fall slightly. These combined changes, which would occur after cleaning, together lower the condenser duty, and it is the difference between the condenser duties in the fouled and cleaned states which constitute the cost of losses attributable to fouling.

The display in Table 6.7 presents the actual heat transfer coefficient calculated from the operating data together with the heat transfer coefficients calculated according to both the ASME and HEI procedures outlined in Chapter 2. From these, the condenser cleanliness or performance factor can be calculated. Further, by modifying the model objectives, the fouling factors can be calculated, which, if applied to the clean condenser model with the presently reduced water flow rate, would cause the compartmental back pressures to assume their current values for the condenser when in its currently fouled state.

Table 6.8 shows the calculated distribution of fouling losses between those attributable to tubesheet fouling and those to tube deposits. These results are obtained by running the clean- condenser model complete with values of the fouling resistances just calculated, but with the water flow rate after cleaning. The difference between the condenser losses calculated for these last two cases as a percentage of the total fouling losses represents the losses which can be attributed to tubesheet fouling (macrofouling). By subtraction, the percentage of the losses attributable to tube-deposit fouling (microfouling) can be obtained.

6.2.4   Carbon Emissions Associated with Fouling Losses

Figure 6.9 includes the calculated hourly fouling losses, which are seen to rise as the fouling resistance increases. By multiplying the fouling losses (MBtu/h) by the cost of fuel ($/MBtu), the equivalent hourly cost in dollars can be calculated.

The two major fuel properties associated with the carbon dioxide emission calculation are carbon content of the fuel (weight) and fuel heating value (HV). Table 6.9 shows typical values for the three major fuels. Now 1 lb carbon produces 3.6644 lb CO_2, and, assuming a boiler combustion efficiency of 95%, the number of pounds of carbon dioxide emissions (CE) per one MBtu change in condenser loss may be calculated from:

$$CE = \frac{3.6644 \times C \times 10^6}{0.95 \times HV}$$
$$= \frac{3.8573 \times 10^6 \times C}{HV} \tag{6.1}$$

To convert the losses due to fouling (MBtu/h) to equivalent carbon dioxide emission due to fouling (lb_m CO_2/h), the data contained in Table 6.9 may be used with Equation (6.1). The last column in Table 6.9 indicates the equivalent carbon emissions per MBtu fouling loss, stated in accordance with accepted IPCC [1995] practice.

Consider the coal-fired unit of the case study of Section 6.2. If the rate of heat loss due to fouling is 34.968 MBtu/h, and assuming that the condenser is cleaned, then the carbon emissions if the unit is operating at this load for 8000 hours per year will be as follows:

$$Carbon\ emissions = \frac{34.968 \times 64.987 \times 8000}{2.06}$$
$$= \underline{9.09}\ million\ pounds\ of\ carbon\ per\ year$$

Table 6.9. Carbon Dioxide Emissions, lb_m CO_2 per MBtu Losses

Fuel	C, lb/lb fuel	HV, Btu/lb_m	lb_m CO_2/ MBtu Loss	lb_m Carbon/ MBtu Loss
Bituminous coal	0.86	13,930	238.1	64.987
Fuel oil	0.863	18,558	179.4	48.950
Natural gas	0.749	25,128	115.0	31.376

Chapter 7

OPTIMIZATION OF CONDENSER CLEANING SCHEDULES

A condenser should be routinely cleaned of debris and fouling in order to maintain low heat rates and/or increase generation capacity. However, like most other aspects of plant operation, condenser maintenance decisions have economic consequences, not only because of the cost of cleaning and the revenue lost while it is being performed, but also because of the cost of fouling losses that may accumulate as a function of the time between consecutive cleanings. If condensers need to be cleaned more than once per year, then the actual months in which cleaning takes place can significantly affect both the revenue and fouling losses as well as the maintenance costs incurred over the operating cycle, which normally extends over 12 to 18 months, but may be much shorter in plants using seawater for cooling.

This chapter describes a calculation method for identifying the month(s) in which a condenser should be cleaned in order to minimize the total cost of losses over an operating period. The selection of the best months is based on

1. An interactive model of the condenser and low-pressure stage of the turbogenerator which is used for condenser performance calculations as discussed in Chapter 5
2. The fouling resistance model for the unit, also developed from the performance calculations and discussed in Section 5.7
3. The historical monthly average cooling-water inlet temperature and generated power for the plant and unit for similar periods in the past

Since condenser operating economics are fundamentally affected by the type and rate of tube fouling, this chapter will first examine some of the causes of fouling and their effects on condenser performance. It also has to be recognized that although linear fouling models can be used for some plants to represent deterioration in condenser performance, the fouling in many plants exhibits nonlinear characteristics, so that a cleaning schedule optimizing technique should allow for both possibilities. And heat transfer is affected directly not only by fouling deposits or surface roughness, but also indirectly by any water velocity variations resulting from tube obstructions. Clearly, the characteristics of the fouling model also play an important part in determining the magnitude of the savings which can be expected from searching for the optimal condenser maintenance schedule.

While infrequent cleaning may not realize the maximum potential savings, a too-frequent cleaning schedule may also not be the most cost-effective. In some cases, the cost of fouling losses tends to decrease with more cleanings, but the lowest-cost solution may still not be a mathematical optimum. Some other criterion has to be chosen: e.g., incremental savings compared with the cost of cleaning, subject to other operating considerations. The *effectiveness* of the selected cleaning method in removing deposits and, hence, minimizing the fouling resistance remaining after cleaning, together with its cost, is also a factor affecting the optimal cleaning schedule. Section 7.3 lists 10 of the many factors which can come into play in evaluating the benefits resulting from cleaning(s). These factors should be taken into account when developing the best cleaning schedule to follow over, say, a 12-month period.

7.1 HISTORICAL PERSPECTIVE

Past practice tended toward conducting plant maintenance principally during the annual plant shutdown. However, reliability-centered maintenance [Smith 1993] and condition-based maintenance [Stanton 1995] have drawn attention to the possibility that a more thorough economic analysis of maintenance and associated costs can offer new opportunities for plant operating costs to be reduced. Among these are techniques for optimizing condenser cleaning schedules, and there have been several attempts to address the uncertainties involved in developing such schedules. Colborn [1923] plotted variations in back pressure over time and suggested that the optimum interval between cleanings could be established by minimizing a mathematical function of cleaning cost, defined as the daily increase in cost of cleaning due to increasing back pressure and time.

In 1981, Ma and Epstein [Ma 1981] discussed falling-rate processes, one form of which is illustrated by fouling model A in Figure 7.1. They indicated how optimum cleaning intervals could be determined by modeling and regressing the mean hourly costs of (losses + cleaning) over time and predicting when this mean hourly cost would approach a minimum. Putman [Putman 1994] applied this technique to condensers subjected to fouling, but found that the forecasting principle functioned best with falling-rate processes and often required that the interest on borrowed money be also taken into account in order to obtain a unique minimum point. Koch, Haines, and Mussalli [Koch 1996] examined the frequency of cleanings and selected the minimum point along a plot of combined cost of losses plus cleaning versus cleaning frequency as the optimizing criterion. However, this approach made assumptions about the water temperature and load profiles over time which may not be valid. More recently, Wolff et al. [1996] outlined how mixed integer linear programming (MILP) can be used to optimize a condenser cleaning schedule but assumed that fouling always increased *linearly* with respect to time. It has been confirmed that the MILP approach can indeed be used for this optimizing purpose, but unfortunately, the method functions only with linear fouling models and cannot be adapted to handle the more common nonlinear models. It was also found that MILP often took more than five minutes before it converged on a solution and always tended to converge on the four cleanings per year for which the MILP matrix had been configured!

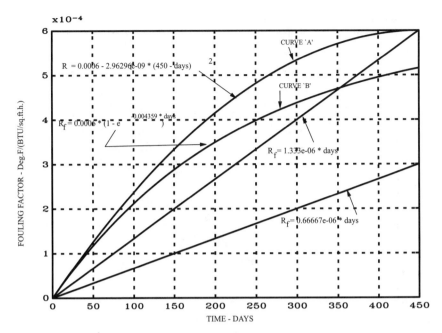

$$\times 10^{-4}$$

FIGURE 7.1. Typical fouling models.

Putman [1997a] explored the work of Wolff et al. further and proposed a method which would allow nonlinear as well as linear fouling models to be accommodated. While the months in which to clean could be identified by the program, the selection depended on the combination of the mean monthly water temperature and load profiles for the unit under study. One finding was that the optimum number of cleanings would sometimes be less than four. It was also found that the *incremental savings* which resulted from an additional cleaning tended to decrease with the total number of cleanings proposed, assuming also that the cleaning method selected would effectively remove the deposits from the condenser on this unit.

To illustrate the point, Figure 7.2 compares the total cost of losses over a 12-month period with no cleanings with that obtained with either one or two cleanings during the same period. There is a significant reduction in losses in month 7 with only one cleaning, and a further reduction is seen if two cleanings were to take place in months 5 and 8. It is clear from this plot that the incremental savings become smaller with each additional cleaning; even so, they can be significant in reducing plant operating costs.

Based on all these considerations, an optimum condenser cleaning schedule program should be able to:

- Handle periods which would correspond to the operating cycle of the plant (weeks or months)
- Accept well-defined historical plant data

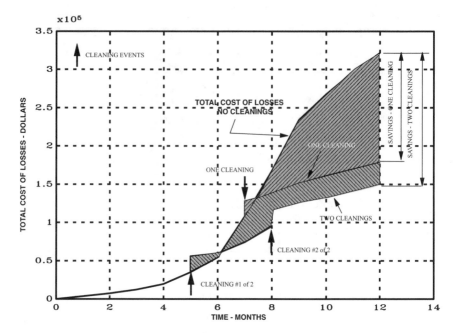

FIGURE 7.2. Cost profiles with multiple cleanings.

• Respond to either nonlinear or linear fouling models
• Provide a solution within the normal human threshold of impatience

The major steps in this optimization analysis include:

1. Construction of a calibrated interactive condenser/low-pressure turbine stage model, generally as described in Chapters 4 and 5.
2. Assembly of sets of historical condenser operating data over the previous period of interest, which, when presented to this model, allows the
   cost of losses and fouling resistance to be determined for each case. The
   fouling model can then be developed from a plot of fouling resistance
   versus time, as indicated in Section 5.7. If on-line condenser performance monitoring is being conducted, then it should not be a difficult
   task to abstract this information from the historical data stored for any
   selected period.
3. Obtaining from the historical data the values of the mean monthly inlet
   water temperature and generated power level, a typical list of which is
   shown in Table 7.1. These parameters, together with the fouling model,
   determine the fouling losses for each subperiod (e.g., month or week)
   covered by the historical averages. If the fuel cost in $/MBtu is known,
   the cost of losses for each subperiod can be calculated by submitting
   the list of mean monthly loads and inlet water temperatures to an off-
   line version of the condenser performance calculation program and cal-

Table 7.1. Mean Values of Cooling-Water Temperature and Generated Power

Month, 1997	Mean CW Inlet Temp.	Mean Generated MW	Mean MW with Shutdowns	Load Factors
Jan	50.97	431.41	227.35	0.527
Feb	50.09	419.38	197.21	0.470
Mar	58.97	433.13	346.39	0.800
Apr	62.95	448.66	279.17	0.622
May	68.80	385.47	335.73	0.871
Jun	74.61	391.87	284.11	0.725
Jul	83.60	365.69	260.01	0.711
Aug	84.48	318.78	285.79	0.896
Sep	81.91	307.65	162.37	0.528
Oct	74.74	359.23	221.62	0.617
Nov	58.48	430.85	335.10	0.778
Dec	51.84	443.44	257.48	0.581

culating the cost of losses after introducing a known fouling resistance. From this the specific cost of losses for the subperiod can be calculated, expressed as dollars per unit of fouling resistance.

4. Optimizing the number of cleaning(s) and identifying the subperiod in which each should occur. This is accomplished using the subperiod profile in step 3 and knowing the fouling model developed in step 2 together with the cost of each cleaning and the residual fouling resistance after cleaning.

7.2 EFFECT OF TUBE FOULING ON CONDENSER PERFORMANCE

Within the Rankine cycle, of the total amount of heat generated in a boiler, some 60% is retained in the vapor exhausted from the low-pressure stage of the turbogenerator. This heat finds its way back into the environment as the condenser removes heat from the exhaust vapor. Unfortunately, this thermal pollution can further increase if the condenser tubes are fouled, and the carbon emissions due to fouling losses should also be taken into account. Table 6.7 showed the relationship between carbon emissions and various types of fuel, for each 1 million Btu reduction in fouling losses. For instance, the carbon emissions into the environment from burning coal are reduced by 65 lb/MBtu of avoided losses. Thus, maintaining a clean condenser can be of benefit not only economically to the plant but also to the environment.

Fouling can have at least two other significant impacts on the economics of unit operation: it always increases unit heat rate, and it can also limit unit generation capacity, especially during periods when cooling-water temperatures are high. The fouling of condenser tubes from contaminants or organisms contained in the cooling water can take many forms:

- Sedimentary or silt formation
- Deposits of organic or inorganic salts
- Microbiological fouling
- Macrobiological fouling

Deposits of sediment entrained with the water can occur when the water velocity falls to the point where the sediment is no longer suspended. To avoid this, water velocities should be maintained as close to their design values as possible. Salts of various kinds can be deposited due to changes in their concentrations, which are affected by the increasing temperature of the water as it passes through the condenser tubes. Salt deposits can consist at least of calcites, sulfites, silicates, or metal oxides of various valences, all of which lower the heat transfer coefficient of the tubes and many of which are very difficult to remove once they have appeared.

Microbiological fouling can occur when organisms rich in metal compounds, e.g., those of manganese, attach themselves to tube surfaces and then produce an impermeable layer. Not only is the heat transfer affected, but many such deposits can also cause underdeposit corrosion, even with stainless steels. This, of course, has a direct effect on tube life. Deposits of this nature are, again, difficult to remove. Macrobiological fouling can occur when mollusks, zebra mussels, and/or similar aquatic organisms pass through the screens and attach themselves either to the tubesheet or to the inside surfaces of tubes. These creatures tend to reduce the water velocity (and so heat transfer), thereby accelerating other types of fouling.

The rate at which each of these fouling mechanisms develops over time is very site-specific and also depends on the chemical treatment methods employed. Thus, a knowledge of the associated fouling model in terms of fouling factor versus time is necessary in order to develop an economic profile of the losses due to fouling. It must be taken into account when developing the cleaning schedule for a condenser over a period of, say, one year.

Figure 7.1 shows a number of different fouling models and the forms of the corresponding equations. The two linear models are obvious, but the nonlinear models of curves A and B can have either a quadratic or an exponential form. The latter curves are also known as falling-rate models, many becoming asymptotic with the progression of time. Note that the equations identified in Figure 7.1 are only examples and the actual mathematical form for a particular unit must be based only on the analysis of the fouling factors calculated from the operating data.

7.3 CONDENSER CLEANING ECONOMICS

As already indicated, the fouling of a condenser not only can affect unit heat rate but may also reduce the maximum power output. Of course, the ingress of air, or the air blanketing of some tubes, can produce effects similar to fouling, but this is outside the scope of the present discussion.

The economic effects of tube fouling can be estimated by quantifying the increased amount of heat which has to be removed from the turbine exhaust vapor as the result of fouling, or quantifying the cost of the power which has to be purchased to compensate for the shortfall in generating capacity. Cleaning the condenser tubes and/or tubesheet almost invariably improves the performance, but when to clean and how often to clean are site-specific decisions.

It must by now be apparent that an extensive knowledge of past condenser behavior is required in order to develop an optimum condenser cleaning schedule for a desired period into the future. The residual thermal resistance which will be experienced after implementing the selected cleaning method should also be known. Following is a list of some 10 major criteria which are either included among the design data for the unit or else need to be developed from historical data; also included are some parameters (e.g. #5 and #10) that must be calculated from the input data by a condenser performance monitoring program:

1. Size of unit
2. Load factor, involving frequency and duration of outages
3. Cost of replacement power, if any, during cleaning operation; or required because generation capacity is presently limited by high back pressure
4. Fuel cost
5. Cooling-water flow rate
6. Inlet water temperature profile during the year
7. Load profile during the year
8. Condenser model
9. Type of fouling deposit
10. Fouling model

Items 1 through 8 are obtained from either the unit design or historical data, and item 9 will be determined from inspection of the deposits. Developing the fouling model (item 10) is probably the most difficult task, but can be facilitated by analyzing data taken from the unit during the period *immediately following a condenser cleaning*, preferably after using metal scrapers. The effectiveness of the method was outlined by Saxon et al. [1996].

The optimizing objective is to so schedule the cleaning operation(s) that the sum of losses plus cost of cleaning(s) over the period is minimized. A measure of the savings can be obtained by comparing the sum of the costs over the period with the costs which would be incurred if no cleaning were to be carried out at all. The cost of each cleaning of a condenser will, of course, depend on the cleaning method to be employed—e.g., metal scrapers, brushes, high-pressure water—together with the number of tubes involved, both of which can be established.

7.4 CONDENSER PERFORMANCE MONITORING

Putman [1995, 1997b] has outlined several methods of modeling the behavior of condensers with single or multiple compartments and with one or two passes.

Since changes in condenser back pressure affect the performance of the turbine, especially the exhaust flow rate and end-point enthalpy, the models are designed to reflect the interactions between the condenser and the low-pressure stage of the turbine.

As has been shown in Chapter 5, such models can be used to estimate the expected condenser duty when the condenser is operating with clean tubes, and to obtain the reference duty for the current operating conditions. The current duty can then be compared with this reference, thus providing an estimate of the current thermal losses due to fouling. The fouling factors in each compartment can also be calculated using this model. The magnitude of the losses is very dependent on the amount of power being generated, as well as the cooling-water inlet temperature, together with its present flow rate. Clearly, the latter can be greatly affected by the presence of any macrobiological fouling, which can obstruct the entry to the condenser tubes or the passages within the tubes themselves. The programs already outlined in Chapter 5 have the following features:

1. Analysis of the expansion line for the current load.
2. Using the expansion line characteristics developed in step 1 to estimate the present condenser duty and cooling-water flow rate for the load and back pressure, both of which are included in the operating data set.
3. Calculation of the duty of a *clean* condenser from the low-pressure stage/condenser model, assuming the unit is operating at the same MW load and cooling-water inlet temperature as in the given data set. Losses due to fouling are calculated from the difference between condenser duty when fouled and the duty when clean.
4. Calculation of heat transfer coefficients and fouling factors for each condenser compartment.

Subsequently, the operating data and results calculated can be postprocessed to:

5. Construct the fouling model based on condenser fouling factors. As mentioned, this model should start with the set taken immediately after the last time the condenser is known to have been cleaned.
6. Calculate the historical mean monthly cooling-water inlet temperatures and generated power levels for those months to be included in the schedule. Typical profiles are shown in Figure 7.3 and listed in Table 7.1.
7. Calculate the specific cost of fouling for each combination of monthly average inlet water temperature and unit load.

7.5 DEVELOPING THE FOULING MODEL

As already indicated, Figure 7.1 shows a number of possible fouling models, each with a unique mathematical formulation. Section 5.7 describes the method of calculating the fouling resistance as an extension of the basic Newton-Raphson

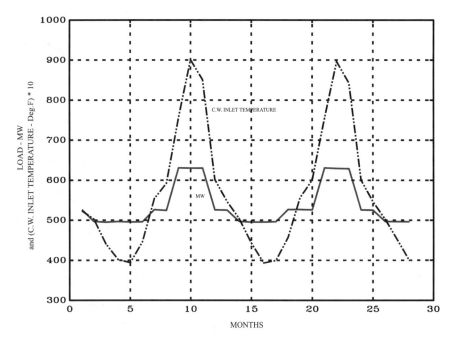

FIGURE 7.3. Average monthly data.

condenser/low-pressure turbine stage model of Chapter 5. The compartmental fouling factors can be calculated for a number of operational data sets and plotted versus time. For the condenser on a particular unit, there will be an appropriate mathematical form for the fouling model. Regression analysis is then used to calculate the coefficients in the selected fouling model using a least-squares fit to the data (see Section A.1 of Appendix A). The validity of such models can be greatly improved if the data sets on which they are based cover the period starting immediately after the condenser tubes have been cleaned, and if the cleaning method which has been found to be the most satisfactory for the unit and type(s) of fouling deposits at the site is used.

7.6 SPECIFIC COST OF FOULING

As has been stated, for a given turbogenerator and its condenser, the condenser duty is a function not only of the physical construction but also of the cooling-water inlet temperature, cooling-water flow rate, and generated power. Thus, it is possible to calculate the losses in MBtu/h due to the presence of known degrees of fouling, given a model of the condenser and turbine low-pressure stage for a unit under study. Figure 7.4 is a plot of such data for the case when the load is held constant but the inlet water temperature is allowed to vary. The data can be fitted by a first-order linear regression having a constant which, for all practical

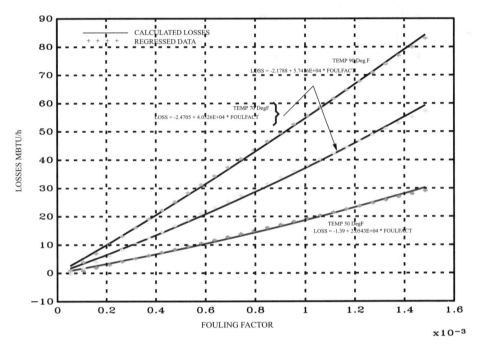

FIGURE 7.4. Fouling factor for constant load with varying inlet water temperature.

purposes, is very close to zero. Similarly, if the water inlet temperature is held constant but the load allowed to vary, the resulting plot in Figure 7.5 shows a first-order fit with the constants, again, being close to zero. That all the curves pass through or close to the origin should not be surprising, since, when the fouling resistance is zero, the fouling losses are, by definition, also zero.

This *essential linearity* of the relationship between losses and fouling factor is an important finding and means that the slope depends on the specific combination of power and water temperature. Once a combination of monthly mean inlet water temperature and power has been established, the duty under equilibrium conditions for a clean condenser can be obtained using the low-pressure stage/condenser model. A fouling resistance of $0.001°F/[Btu/(ft^2 \cdot h)]$ can then be applied and the new duty calculated when the model has reestablished equilibrium for this combination of cooling-water temperature and power with the condenser fouled. The difference between these two condenser duties is the fouling loss (FL) in MBtu/h for a fouling resistance of $0.001°F/[Btu/(ft^2 \cdot h)]$. Thus, if the *specific cost of fouling* is defined as the cost of losses per $°F/[Btu/(ft^2 \cdot h)]$, at the given inlet water temperature and generated power, per unit of fouling resistance, then the specific monthly cost of fouling (SC) for each month under analysis can be calculated from:

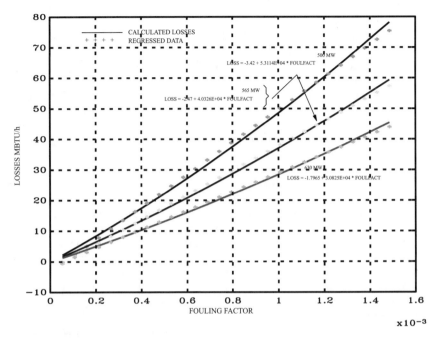

FIGURE 7.5. Fouling factor for varying load with constant inlet water temperature.

$$SC = \frac{FL}{0.001} \times 24D_{month} \times \$/MBtu \qquad (7.1)$$

where D_{month} is the number of days in the month and $\$/MBtu$ is the fuel cost.

Given that the relationship between losses and fouling is essentially linear and passes through the origin, multiplying the fouling resistances calculated from the fouling model versus time by the specific cost of losses provides a direct measure of the cost of losses due to fouling.

7.7 OPTIMIZATION PROGRAM STRUCTURE

The structure of the cleaning schedule optimization program is very simple and is based on a binary search of all the possibilities when cleaning once, twice, three times, or four times in a 12-month period, starting from a selected month. A permissive flag can be set for each month to indicate whether cleaning during that month is allowed. In this way, it is possible to avoid assigning shutdowns during the hot summer months when generation can command a premium.

The input data required by this program consists of:

- Specific cost of fouling losses for each month.
- The cost of each cleaning of the condenser.
- The residual fouling factor experienced after a cleaning.
- Mathematical form of fouling model and constants for each term in the model. From this, the fouling resistance can be calculated for each month after a cleaning has taken place, either in actuality or within the simulation forming part of the cleaning schedule optimization program.
- Those months in which cleaning is not permitted, when the analysis is applied to data extending over one year or more.

To calculate the total cost of losses plus cleaning if there is to be only one cleaning in the 12-month period, an arbitrarily high value is first assigned to the minimum annual cost. The program then steps through each month; calculates the set of fouling resistances for each month in the period both before and after the assumed cleaning, and calculates the cost of losses for the fouling resistance profile corresponding to the selected cleaning month. This is compared with the previously established minimum, and if the new value is lower, the minimum cost value is updated and the corresponding cleaning month identified. At the end of the program, that month for which the calculated cost of losses is the smallest is the optimum month in which to clean when only one cleaning is anticipated during the period.

To test whether there should be two cleanings in a 12-month period, a rule is introduced to ensure that at least one month must separate two consecutive cleanings. Thus, the initial assignments are month 1 and month 3 of the period, the monthly resistance profile for this combination being calculated together with the total cost of losses. With month 1 continuing to be assigned, the other month is moved from 3 to 12 in incremental steps. The fouling resistance profile for each pair is calculated at each step and for the whole period, together with the corresponding total cost of losses. That pair for which the lowest total cost of losses was calculated is the optimum combination for the two-cleanings case. A similar procedure is followed for both the three-cleanings and four-cleanings cases, all lowest-cost cases being selected and displayed.

This program can be executed on a personal computer with a Pentium processor and takes a total of 5 seconds to run to completion as well as display the results for each case. By comparison, if mixed integer linear programming is used as the optimization method (a method only valid for use with a linear fouling model), then a complete analysis for four cleanings can take up to 5 minutes to run to completion on the same personal computer.

7.8 TYPICAL OPTIMUM CLEANING SCHEDULE

In the following, data taken from a 550-MW unit in Tennessee has been used to illustrate how an optimum cleaning schedule can be developed. However, it must be pointed out that the magnitude of the losses and the tabulated results are unique to this unit. For other units with different operating parameter profiles and unit load factors, as well as different fouling models, it is possible that

Table 7.2. MW from Column 3, Table 7.1

MW	CW Temp, In	Duty, Clean	Duty, Fouled	Fouling Losses	Days in Month	Specific Cost	Fouling Resistance	Cost Without Cleaning
431.41	50.97	1740.28	1767.43	27.15	31	2019.96	2.5000E-05	580.74
419.38	50.09	1691.22	1716.96	25.74	28	1729.73	2.2000E-04	4,376.21
433.13	58.97	1764.12	1795.09	30.96	31	2303.42	5.0000E-04	13,244.69
448.66	62.95	1835.21	1869.92	34.71	30	2499.12	7.5000E-04	21,554.91
385.47	68.80	1598.13	1629.56	31.43	31	2338.39	9.0000E-04	24,202.36
391.87	74.61	1638.88	1674.79	35.91	30	2585.52	9.8000E-04	29,138.81
365.69	83.60	1560.22	1599.72	39.51	31	2939.54	1.0400E-03	35,156.95
318.78	84.48	1373.71	1407.65	33.94	31	2525.14	1.0400E-03	30,200.63
307.65	81.91	1321.40	1352.10	30.70	30	2210.40	1.0400E-03	26,436.38
359.23	74.74	1508.63	1540.79	32.16	31	2392.70	1.0400E-03	28,616.74
430.85	58.48	1753.99	1784.49	30.50	30	2196.00	1.0400E-03	26,264.16
443.44	51.84	1789.46	1818.04	28.58	31	2126.35	1.0400E-03	25,431.17

Yearly cost of losses—no cleanings $265,203.74

another optimum cleaning schedule will be developed, while the operating and maintenance cost reductions may be greater or less than those indicated for this unit. For instance, the operating cost reductions on a 50-MW unit will be significantly smaller than those obtained with an 800-MW unit. Thus, the schedule-optimizing method outlined in this section is probably of more interest than the actual values generated.

Once the fouling model and the monthly specific cost of fouling have been computed, a data table can be constructed as shown in Table 7.2. This shows how Table 7.1 can be extended to include the fouling model and the specific cost of losses for each combination of mean monthly cooling-water inlet temperature and generated power. The last column shows the cost without cleaning for each month, assuming the fouling model will be faithfully followed over the whole period. The sum of these monthly losses is the total cost of losses over the period if no cleanings were to be carried out.

All of this data is then presented to the optimization program, the results from which are included in Table 7.3. The first line below the heading for each set of optimum results shows a '1' for each month when cleaning is to occur. The second and third lines contain the fouling factors assumed for each month, a factor of 0.25 being shown for the month(s) when cleaning is to occur. The right-hand column indicates the cost of losses (including cost of cleaning(s)) for one, two, three, or four cleanings per year if the mean monthly generated power is that shown in column 3 of Table 7.1. In this case, a condenser cleaning cost of $13,500 has been assumed. Note that with this fouling model there is no mathematical optimum, only a minimum cost. If the cost of losses without cleaning were $265,203, then cleaning three times during the year would save $132,571. Cleaning a fourth time would save only an additional $9111. Thus, the best choice could be considered to be three cleanings during the year.

Table 7.3. Optimum Schedule-MW in Column 3, Table 7.1

	J	F	M	A	M	J	J	A	S	O	N	D	Months				Cost of losses
Optimum for one cleaning																	
	0	0	0	0	0	1	0	0	0	0	0	0	0	0	0	0	$193,969
	2.8791	5.4366	7.3791	5.4366	7.3791	8.7872	9.7412	8.7872	9.7412	9.7412	.2500						
	2.8791	5.4366	7.3791	5.4366	7.3791	8.7872	9.7412	8.7872	7.3791	9.7412	10.3217						
Optimum for two cleanings																	
	0	0	0	1	0	0	0	1	0	0	0	0	4	8	0	0	$157,214
	2.8791	5.4366	7.3791	5.4366	7.3791	7.3791	2.8791	.2500	2.8791	2.8791	5.4366	5.4366					
	7.3791	7.3791	7.3791	5.4366	2.8791	2.8791	7.3791	7.3791	7.3791	7.3791	8.7872	8.7872					
Optimum for three cleanings																	
	0	0	0	1	0	0	1	0	0	1	0	0	4	7	10	0	$132,633
	2.8791	5.4366	7.3791	.2500	2.8791	7.3791	.2500	2.8791	2.8791	2.8791	5.4366	5.4366					
	.2500	5.4366	5.4366	.2500	5.4366	5.4366	.2500	2.8791	2.8791	2.8791	5.4366	5.4366					
Optimum for four cleanings																	
	0	0	1	0	1	0	1	0	0	1	0	0	3	5	7	10	$123,522
	2.8791	5.4366	.2500	5.4366	.2500	2.8791	.2500	2.8791	2.8791	2.8791	2.8791	2.8791					
	.2500	2.8791	5.4366	2.8791	5.4366	5.4366	5.4366	2.8791	2.8791	2.8791	5.4366	5.4366					

Table 7.4. Return-on-Investment Analysis, Mean Loads

	1	2	3	4	5
		Total			Return
	Cost of	Cost of	Cost of	Annual	on
	Losses	Cleaning	Fouling	Savings	Cleaning
No cleanings	265,204	0	265,204	0	0
One cleaning	180,469	13,500	193,969	71,235	84,735
Two cleanings	130,214	27,000	157,214	107,990	50,255
Three cleanings	92,133	40,500	132,633	132,571	38,081
Four cleanings	69,522	54,000	123,522	141,682	22,611

Table 7.4 includes the results tabulated in Table 7.3 but also shows the returns for each incremental cleaning when there are no outages during the year. The annual savings shown in column 4 are obtained by subtracting the total of costs of fouling with n cleanings shown in column 3 from the cost of fouling with no cleanings, i.e., \$265,204, contained in the first row of column 3. The incremental return on cleaning n, shown in column 5, is obtained by subtracting the cost of losses shown in column 1 for cleaning $n + 1$ from the cost of losses shown in the same column for cleaning n.

Table 7.5 shows similar information if the unit is taken out of service for periods corresponding to the load factor shown in the fifth column of Table 7.1. Using the specific costs shown in Table 7.2 corrected for outages, the yearly cost of losses with no cleanings is shown to be \$190,680. With three cleanings during the year, the cost of losses is reduced to \$104,681, a net savings of \$85,999. This indicates that the cost reduction after cleanings on a unit operating for most of the year close to full load will be far greater than the cost reduction on a unit operating at part loads for long periods, and perhaps frequently taken out of service. This suggests that load factor has a major influence on the magnitude of the net reduction in fouling costs.

Table 7.5. Return-on-Investment Analysis: Mean Loads × Load Factor

	1	2	3	4	5
		Total			Return
	Cost of	Cost of	Cost of	Annual	on
	Losses	Cleaning	Fouling	Savings	Cleaning
No cleanings	190,680	0	190,680	0	0
One cleaning	126,334	13,500	139,834	50,846	64,346
Two cleanings	88,194	27,000	115,194	75,486	38,140
Three cleanings	64,181	40,500	104,681	85,999	24,013
Four cleanings	45,182	54,000	99,182	91,498	18,999

Note that, for a given cleaning frequency, the months when cleanings should occur have been developed from an optimization program which forces these schedules to be optimum at the selected frequency. In this case, however, increasing the cleaning frequency did not provide a truly mathematical optimum, even though the savings do increase with the number of cleanings. Unfortunately, the *incremental* savings become progressively smaller.

Thus, rather than searching only for the mathematically optimum schedule, a more useful criterion may be the "return" obtained from a cleaning, "return" being defined as the reduction in cost of losses for an additional cleaning of the condenser. If the minimum return on a cleaning investment is assumed to be 2:1, then the last column of Table 7.4 indicates that three cleanings per year would be chosen if there were no outages. This choice predicts a net annual savings of $132,571. However, if outages are to be taken into account, then, according to the last column of Table 7.5, two cleanings would be chosen, resulting in an annual savings of $75,486.

None of these calculations has taken account of any replacement power cost. Where power has to be purchased, its cost should be added to the cost of cleaning.

Chapter 8

ECONOMIC IMPORTANCE OF MANAGED CONDENSER MAINTENANCE

Of the total heat consumed in the boiler within the Rankine cycle, some 60% is retained in the vapor exhausted from the low-pressure stage of the turbogenerator and only 40% is converted to power. Thus the condenser acts as a heat sink, while also keeping the back pressure as low as possible, and so maximizing the amount of energy converted to useful work. The condenser also recovers the vapor in the form of condensate, which can then be recycled without requiring any costly form of water treatment.

As already indicated, condenser performance deteriorates mainly because of progressive fouling of internal tube surfaces or the tubesheet, and/or because of inadequate removal of air or other non-condensibles from the condenser shell. All of these effects tend to lower the heat transfer capacity of the tube bundles and can reduce the performance of the condenser and so of the boiler/turbogenerator unit. However, careful management of condenser maintenance can contribute significantly to improving a unit's economic performance. While the set of economic objectives chosen for a particular plant may vary, they will fall into the following categories:

- Minimizing boiler/turbogenerator unit heat rate
- Maximizing generation capacity
- Minimizing losses due to fouling
- Minimizing carbon and other emissions
- Maximizing availability

8.1 MINIMIZING UNIT HEAT RATE

In the United States, a unit's heat rate is stated in Btu/kWh and reflects the total amount of heat required per hour to generate the steam and reheat it, divided by the number of kilowatt-hours generated. Heat rate is not a fixed number but varies between a high value for operation under partial-load conditions and a minimum value at the MW load corresponding to the maximum continuous rating (MCR) of the unit. However, at loads above the MCR, it is not uncommon for the heat rate to rise slightly. Thus, from its definition, the heat rate of a turbogenerator unit is a measure of the total amount of energy supplied to the unit in order to generate one kilowatt-hour of electricity, or a measure of the total input/output efficiency. To convert input/output efficiency to the normal output/input efficiency, consider a unit with a heat rate of 10,000 Btu/kWh. The relationship between electrical energy and the equivalent thermal energy is:

$$1 kWh = 3412.13 Btu$$

from which the output/input efficiency of a unit (Eff_{unit}) with a heat rate of 10,000 Btu/kWh is:

$$Eff_{Unit} = \frac{3412.13}{10,000} = 34.12\%$$

Clearly, the lower the heat rate, the higher the output/input efficiency.

The *unit* heat rate is the total heat input required to generate the steam, divided by kWh output [ASME 1988], with all boiler losses being included, whereas *turbogenerator* heat rate is the total heat acquired only by the steam itself (both main and reheat steam) divided by kWh output. Furthermore, the turbogenerator heat rate may be stated as the *gross heat rate*, with the total power delivered by the generator as the denominator. Alternatively, it may be stated as *net heat rate*, in which only the net power delivered to the network forms the denominator, with any auxiliary power consumed by unit auxiliary equipment subtracted from the total generator load. Thus, for a turbogenerator, the net heat rate is always higher than the gross heat rate, which is lower than the equivalent unit heat rate. Since the economics of operating the unit as a whole is of particular concern with regard to the condenser, the term *heat rate* in the following text always applies to unit heat rate.

Some of the reasons the unit heat rate may appear higher than design for a given load are as follows:

- Boiler operating at a lower than design combustion efficiency
- Boiler steam temperature(s) lower than design
- Boiler flue-gas passes fouled
- Feed heaters out of service, causing boiler water inlet temperature to be low
- Damaged turbine blades
- Excessively high turbine-gland leakage
- Steam-dump valve leaking
- Fouling of the condenser tubes
- Fouling of the tubesheet
- Inadequate removal of air or other non-condensibles
- Excessive air inleakage to vacuum components and piping
- Condensate temperature below shell saturation temperature (subcooled)
- Waterboxes not running full
- Insufficient circulating water pumps in operation
- High cooling-water inlet temperature
- Inaccurate instrumentation

As explained in Section 1.1.3, condenser fouling causes the exhaust back pressure to rise, so that the associated increase in exhaust enthalpy reduces the amount of thermal energy in the vapor which is converted to power. In a fossil plant, an increase in the back pressure can also result in an increase in the boiler steam flow (or turbine throttle flow) required to generate a given amount of power, and thus more heat. Regardless of whether it is a nuclear or a fossil unit which is under investigation, a deterioration in condenser performance almost always results in an increase in unit heat rate and the associated costs involved.

8.2 MAXIMIZING GENERATION CAPACITY

In addition to increasing unit heat rate, a falling off in condenser performance can also limit generating capacity. The turbine manufacturer normally limits the maximum back pressure so as to avoid problems with the low-pressure turbine blading. Whenever the back pressure in any condenser compartment tends to rise above this upper limit (typically 5.0 in-Hg, but possibly as high as 5.5 in-Hg), the load on the unit must be reduced. This results directly in a loss of revenue, since fewer MWh's are now being delivered to the grid. The costs can grow if replacement power has to be purchased, especially at short notice. In a deregulated environment, the spot price of power has been known to rise to as high as $7500/MWh, compared with the normal range of $25–35/MWh.

With fossil plants, capacity limitations are seldom of concern during the winter months. However, during summer peak-load periods, it is quite usual for generation capacity to become limited, as a high circulating-water inlet temperature causes a falling off in condenser performance and resulting higher back pressures. This can occur whether the cooling water is drawn from a river, a lake, the sea, or a cooling-tower pond.

In nuclear plants, the conditions specified in the operating license determine the amount of heat available for generating steam. They are normally baseloaded, and operated to maximize the power this heat can generate. Since the amount of steam available, and the conditions under which it is delivered, are a function of the reactor, the conditions in the condenser have a great influence on the power the unit can generate and the revenue flow which can be expected. Thus, some nuclear plants have made condenser maintenance a major vehicle for revenue improvement. DeMoss [1995], Stiemsma et al. [1994], and Saxon [Saxon and Putman 1996] show how 25 MW of capacity was recovered on a unit at the Clinton plant of Illinois Power Company and 6 MW at the Peach Bottom station of PECO Energy Company. Such increases in unit capacity quite often result from improved condenser maintenance techniques and practices, as experienced in these and other cases.

8.3 MINIMIZING LOSSES DUE TO FOULING

In Section 8.1, a number of factors which can affect unit heat rate were mentioned, each of which has its own economic impact on operating costs. Deviations from design for some are often displayed to the unit operator as a set of controllable losses; deviation in condenser back pressure is often among these. *Any* falling off in condenser performance will make its own contribution to operating costs (or controllable losses) through an observed increase in unit heat rate, the magnitude varying with load, degree of fouling, concentration of non-condensibles and various other factors.

Among the set of controllable losses presented to the operator is steam temperature deviation, which can often be reduced quickly by making appropriate adjustments to the boiler control system. On the other hand, the only method for quickly reducing back pressure deviation is if it is due to low cooling water flow and additional pumps are available to be switched on. Otherwise, since fouling of the condenser usually occurs slowly, the inclusion of back-pressure deviation among the set of controllable losses provides only a warning of condenser performance deviation: correction can be carried out only through maintenance and often requires a unit outage.

The *controllable* losses traditionally assigned to back-pressure deviation are obtained from the curves provided by the turbine manufacturer, which show the heat-rate deviations corresponding to back pressures at various loads, generally as shown in Figures 2.1 through 2.3. Note that these losses are costs associated with *deviations* from design, due to a number of possible causes including air ingress, off-design cooling-water flow rate and temperature conditions, and fouling. For any given condenser design specification, generated power, and cooling-water inlet temperature and flow rate, there will be *unavoidable* losses: a certain amount of latent heat will be transferred to the cooling water even if the condenser is clean, and it should be noted that both this heat and the associated deviation in back pressure may be different from the values contained in the condenser design data sheet. If the condenser becomes fouled, clearly these particular losses will increase. Thus, the degree of condenser fouling affects only the cost equivalent of the *difference* between the heats of condensation removed by a fouled and a clean condenser (avoidable losses). For this reason, not all of the controllable losses defined as a function of back-pressure deviation may be reducible in practice. In order for sound economic decisions to be made, the losses to be assigned to fouling, air ingress, or insufficient flow must be individually quantified.

The condenser performance models outlined in Chapter 5 allow the fouling resistance of any deposits to be established and the hourly cost of losses due to fouling to be quantified at any particular point in time. Chapter 7 outlined how this archived data can be used (1) as the basis for constructing fouling models for a given unit, (2) to establish the average monthly cooling-water inlet temperatures and generated power experienced for this unit, and (3) to optimize the months in which the unit should be cleaned if the total cost of losses over an operating period are to be minimized.

8.4 MINIMIZING CARBON AND OTHER EMISSIONS

In fossil plants, the heat discharged into the environment increases when the performance of the condenser deteriorates due to fouling, air ingress, or increased concentrations of non-condensible gases. These increased losses cause more fuel to be consumed in order to satisfy power generation requirements. In the process, emissions of both sulfur and carbon dioxides increase, together with any associated NO_x discharges. These emissions must be avoided [IPCC 1995]. Thus, poor condenser performance not only increases the amount of heat consumed unnecessarily but also has a negative environmental impact.

This concern for fuel economy and conservation is not new. Those engaged in the utility industry during World War II and immediately after were constantly involved in minimizing the heat consumed in order to maximize the generation of power. The combustion control industry, which was founded in 1916, made further improvements in the efficiency of boiler operation as well as the closer control of the temperature and pressure of the steam supplied to the turbogenerators. Even though the early mechanical control systems were based on hydraulic and pneumatic principles, they were still effective. The subsequent conversion to electronic and, eventually, digital technology made still further improvements. Given existing combustion processes and equipment, the major constraint on even greater improvements has been the limits imposed by the laws of thermodynamics, which, unfortunately, any amount of legislation cannot change.

Engineers have always considered conservation of energy a moral imperative: their hallmark of outstanding engineering performance is successful improvement in process efficiency, fundamentally an energy conservation concept. Unfortunately, this ideal is not always met in practice and penalties for non-performance have been instituted: first, by imposing economic penalties on emissions of sulfur dioxide and now, if the proponents of global warming have their way, carbon dioxide as well. In the case of SO_2, the federal government has stipulated that emissions greater than a set amount from each plant will be penalized, while emissions below these targets can provide a credit. These SO_2 credits can be traded, so that those whose emissions increase unduly can, by buying these credits, avoid paying additional penalties for their violations.

In December 1997 the U.S. Government signed the Kyoto Protocol in which it agreed to reduce the amount of carbon dioxide emissions to 7% below their level 1990 levels, by the year 2012. Thus, one aftermath of the Kyoto Protocol has been an intense international discussion regarding instituting a similar system of credits for carbon dioxide emissions, allowing these to be traded internationally rather than merely within the United States. Since the United States is unlikely to meet its Kyoto obligations within the time frame allowed, it will be possible to purchase credits from, for example, Russia, whose industry is currently discharging much less carbon dioxide than in 1990. Carbon dioxide emissions under the treaty will carry economic penalties, estimated to be between \$14 and \$348 per ton of "greenhouse gases" emitted [VanDoren 1999], which, if instituted, will have to be taken into account when developing estimates of the savings resulting from avoiding unnecessary emissions due to deterioration in condenser performance. (Table 6.7 contained data used to estimate the carbon emissions saved for each MBtu reduction in avoidable losses.)

8.5 MAXIMIZING AVAILABILITY

Operating a turbogenerator unit at its maximum efficiency and capacity can, clearly, help to generate increased revenue, but ensuring that it is also available to deliver power to the transmission system whenever needed allows revenue to be maximized. Maximizing condenser efficiency entails:

- Maintaining the tubes in good condition, free from cooling-water leaks into the condensate, and free from deposits when they will tend to affect generation capacity as well as heat rate
- Paying attention not only to the main tube bundles but also the air-removal section, which, if fouled, might prevent adequate removal of the non-condensibles
- Also keeping free from excessive air inleakage those parts of the turbo-generator operating at subatmospheric pressures, because the air tends to accumulate in the condenser, negatively affecting heat transfer and increasing the duty on the air-removal system

Monitoring the rate of deterioration in any of these criteria can only improve the proper planning of maintenance activities. In addition, when optimizing a condenser cleaning schedule, those months in which cleaning must not occur must also be identified, as mentioned in Section 7.7. For other maintenance criteria, the principles of reliability-centered maintenance [Smith 1993] should be employed. Thus, maintenance should not be confined to scheduled annual shut-downs, which may be too late to prevent unscheduled outages.

Chapter 9

UNIT OPERATOR DIAGNOSTICS

Chapter 8 described some of the economic motivations which can only stimulate a keen and continuous interest in the on-line performance of the condenser and its auxiliary equipment. However, the criteria against which the available information is to be interpreted vary with the load on the unit as well as with changes in ambient conditions. This chapter outlines some approaches to interpreting this information to establish whether the condenser is, in fact, performing below its capability. While there are a number of rules which can be followed, problems tend to be very unit-specific, especially when the condenser has more than one compartment and each is operating at a different pressure as well as a different cooling-water temperature. (In this chapter, the diagnosis of possible faults will focus on the behavior of a single-compartment condenser; complications introduced by more than one compartment will be discussed separately.)

Some of the instruments usually available for condenser performance monitoring were discussed in Section 3.3. Other information commonly available includes pump motor currents, the status of motor breakers, and the concentration of non-condensible gases in the condenser off-gas. Of course, it is always possible for key instruments to lose their calibration. When this occurs, the set of information can become very misleading, often causing valuable time to be wasted in solving what turns out to be a nonproblem. With this in mind, the following will assume that the available data is accurate and that rational conclusions can be drawn from it.

Before attempting either to verify that the condenser is performing in accordance with design or to diagnose whether its condition has changed, the analyst must have available the key design information, not only for the condenser itself but also for the turbogenerator, this being abstracted from the thermal kits provided by the turbine builder. The configuration of the condenser and its water flow path must also be kept in mind. Sometimes condensate chemistry can be an indicator of system malfunction. The behavior of the air-removal system is another important factor to be considered.

Improving on design conditions is outside the scope of this discussion. Such efforts can require extensive computerized fluid dynamic (CFD) modeling of the vapor flow through the condenser, a critical analysis of the design of the air-removal system, and a study of the possibility that some tubes may have to be removed to improve laning. Using help from CFD specialists, some successes [Frisina 1989, Rhodes 1998] in improving the capacity of a condenser have been reported, but these are not techniques which operators will have at their disposal. Unfortunately, a successful CFD analysis does not always lead to a problem solution. Tucci and Bell [Tucci 1999] report on a case in which CFD analysis of a condenser with air binding problems identified the affected groups of tubes but found there was no cost-effective way to eliminate the fundamental problem.

A distinction must also be made between problems which become more obvious over time and problems which emerge suddenly, e.g., a sharp increase in off-gas air concentration as a result of a leak developing in the turbine casing. Many condenser problems have also been known to appear anew immediately after a plant outage.

9.1 REFERENCE CONDITIONS

In the past, attempts to diagnose condenser performance have normally been conducted under full-load conditions, where deviations from the initial design performance are easier to identify. These are also the conditions on which ASME [1998] and HEI [1995] acceptance test standards were based. However, plants are increasingly being operated in a cycling mode so that extended operation under full-load conditions cannot be assured. In any case, a diagnostic procedure must begin with a clear idea of how the condenser ought to be performing under *current* conditions of load and cooling-water inlet temperature and flow rate. Chapters 3 through 5 discussed how interactive models of the condenser/low-pressure turbine stage may be used to establish the operating conditions throughout the system, given the current load and cooling-water conditions, and assuming the condenser to be operating in a clean state. This set of conditions can become the reference to compare with present measurements.

Table 9.1 shows a summary display from a condenser performance monitoring program for a single-compartment condenser, containing the current operating conditions as well as the equivalent set of conditions calculated assuming the condenser to be in its clean state. Note that only the generator load, cooling-water flow rate, and cooling-water inlet temperature for the clean state are the same as in the set of present conditions. Note also that, if the condenser is first in a fouled state and is then cleaned, then the back pressure and outlet water temperature as well as condenser duty will all be less than their present values. The losses due to fouling are the difference between the present duty and the duty with a clean condenser. This avoidable loss can be directly converted to a dollar value by multiplying by the cost of fuel in $/MBtu. Thus, the cost of losses will vary with not only the degree of fouling but also the cost of fuel, the magnitude of the costs drawing attention to the penalty incurred by fouling.

Even under full-load conditions, the reference set of operating parameters may be different from design. For instance, the design back pressure given in the condenser design data sheet is that associated at full load with the stated cooling-water inlet temperature, this usually being the highest temperature which will be experienced by the unit during the summer. It will therefore be readily understood that a lower water inlet temperature will, even with a clean condenser, cause the back pressure to fall below its design value at full load. Similarly, if one circulating water pump is shut down, causing the flow to fall, the cooling-water inlet temperature will determine whether the back pressure will then rise above design.

Table 9.1. Monitoring Results for a Single-Compartment Condenser

05-03-1999 10.32	CONDENSER PERFORMANCE MONITORING PROGRAM RESULTS

Generator Load	=	568.000	MW	
Fouling Losses	=	212.149	MBtu/h	
Fouling Resistance	=	.00002967	Deg.F/Btu	

		PRESENT CONDITIONS	CLEAN CONDITIONS	
Condenser Duty	=	2376.710	2164.562	MBtu/h
Outlet Temperature	=	93.750	91.630	Deg.F
Steam Temperature	=	100.400	97.385	Deg.F
Back Pressure	=	1.95680	1.78638	in-Hg
Exhaust Flow	=	2451300.000	2302649.000	lb/h
CW Flow	=	200000.000	200000.000	GPM
Inlet Temperature	=	70.000	70.000	Deg.F
ASME U Coefficient	=	782.933	801.553	Btu/sq.ft.h

Other reference data, such as design concentration of air in the condenser off-gas, can be obtained from the HEI standard [1995]; while the design total vapor flow as a function of exhaust flow and number of compartments can be obtained from the ASME standard [1998]. The expected concentration of dissolved oxygen in the condensate, the design cooling-water pressure drop, and the value of the terminal temperature difference versus generated power should all be available from within the design data set.

9.2 SOME CAUSES OF DEVIATE CONDENSER PERFORMANCE

A comparison of the data sets contained in the two columns of Table 9.1 is one place to look for an indication that some change in condenser performance has occurred. Other indications would be increases in condensate dissolved oxygen concentration or salinity. The accuracy and location of the measuring instruments must also be taken into account. Figure 9.1 shows the general location of the instrumentation set usually provided for the condenser. Note that, because of the possibility of stratification, cooling-water outlet temperatures should be measured at several points close to the outlet waterbox and their values averaged. Similarly, several instruments should be provided to measure compartmental back pressures and their values also averaged. Clearly, not only will the signals provide direct information, but differential pressures and temperatures can be computed from them.

To the operator, the most obvious and sometimes dramatic symptom that conditions internal to the condenser may have changed is that the back pressure is

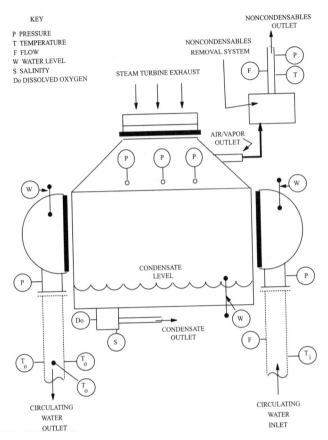

FIGURE 9.1. Location of condenser instrumentation.

higher than expected. The judgment that further attention is called for may be intuitive, or it may be based on the reference value calculated for current conditions. The need for attention may also be unequivocal if, say, the pressure is approaching the alarm point, commonly 5.0 or 5.5 in-Hg. A change seen in back pressure can be the result of one or more of several factors, which can be grouped as to whether the factor affects the circulating-water side or the shell side of the condenser (or both, if the problem is a tube leak).

9.2.1 Circulating-Water Side

Low Water Flow Rate

Water flow rate is rarely measured directly using instrumentation; Chapter 4 described a method of estimating it using thermal kit data. A low cooling-water flow rate could occur because the water passages have become obstructed (by debris or fouling), or because some pumps have been switched off and/or their impellers have become worn. It should also be remembered that control panel instrumentation can sometimes be misleading. At one site the control panel lights showed that a pump was running, yet a physical examination showed that the shaft of that pump was not rotating!

Cooling-Water Pressure Drop High

A large pressure drop will be associated with a reduced water flow rate and is a sign that the tubes or tubesheet have become fouled.

Waterboxes Not Running Full

If the inlet waterbox is not running full, then some of the upper tubes will become exposed and no water will flow through them. Whether or not this is the case can be confirmed by checking the level gage mounted on the waterbox, or by checking if water is discharged on opening a vent valve located on the top of the box. A low waterbox level can sometimes be associated with a unit on which the cooling-water pumps have to be primed before being started.

Occasionally the level in the *outlet* waterbox is found to be low. This is due to some air inleakage, which causes the vacuum to be broken. When this happens, the water flow rate can fall because of the reduced hydraulic head [Fromberg 1997].

Fouling of Inner Surfaces of Condenser Tubes

Fouling of tube inner surfaces is one of the most common causes for the back pressure to rise above its expected value. The operator who suspects that such fouling has occurred may first look at the record: it will be known how long it has been since the condenser was last cleaned, how it was cleaned, and whether performance returned to normal after the last cleaning. The *rate* of fouling will also be known from past experience. Another way to confirm tube fouling is to compare the present cleanliness factor with that expected according to design for the same load (see Section 2.4). And tube surface fouling will affect the terminal temperature difference, but it may not have sufficient effect on the flow rate to confirm the diagnosis, although the flow rate can be affected by serious tubesheet fouling.

Fouling may also be confirmed by using the values of the U coefficients calculated for present and clean conditions (U_{fouled} and U_{clean}, respectively; see Table 9.1) and calculating the fouling resistance as follows:

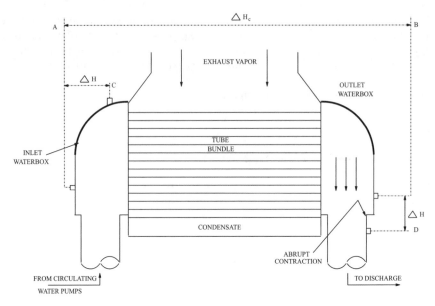

FIGURE 9.2. Fouling detection by differential pressure comparison.

$$R_f = \frac{1}{U_{\text{fouled}}} - \frac{1}{U_{\text{clean}}} \tag{9.1}$$

Yet another approach is one developed by the Tennessee Valley Authority [March and Almquist 1987], in which the pressure drop between the inlet and outlet waterboxes (ΔH_c) is compared with the pressure drop ΔH measured across an abrupt contraction in the outlet waterbox (see Figure 9.2). The pressure drop across a clean condenser is a function of the square of the flow rate Q multiplied by a constant K_c, the value of which can be obtained from operating data:

$$\Delta H_c = K_c Q^2 \tag{9.2}$$

Similarly, the pressure drop across an abrupt contraction in the outlet waterbox piping follows the law

$$\Delta H = K Q^2 \tag{9.3}$$

The values of both K and K_c can be calibrated from the values of ΔH and ΔH_c taken at the same time and when the water flow through the condenser has the design value at full load. Another calibration point can be obtained by observing the values of ΔH_c and ΔH after one pump has been turned off. Equation (9.3) can also be used when the pressure difference is measured across a diverter or bend in the inlet piping or waterbox. In either case, fouling will have little effect on the relationship of Equation (9.3). If the calibration data is taken when the condenser is clean, it will be found that:

$$\frac{K_c}{K} = \frac{\Delta H_c}{\Delta H} = c \qquad (9.4)$$

Subsequently, the value of $\Delta H_c/\Delta H$ is tracked against the value of the constant c in Equation (9.4), deviations above the basic value indicating that fouling is probably occurring.

Low Cleanliness Factor

If the current cleanliness factor is less than should be expected from design, it can also be an indicator of fouling or even of air ingress. Since cleanliness factor has been shown to vary with load [Putman 1999], care must be taken that an appropriate comparison is being made (see Section 2.4).

Pump Motor Current

A high pump motor current reading can be a sign that the pump is deteriorating or an indication of a high condenser water pressure drop, possibly due to tube and/or tubesheet fouling. Either cause can signify substandard condenser performance.

Subcooling

Subcooling is considered detrimental to unit heat rate and should be minimized. However, the degree to which observed subcooling is *avoidable* is not readily determined. Subcooling is said to occur when the temperature of the condensate is below the saturation temperature corresponding to the present shell pressure. However, as pointed out in Section 1.3, the temperature of the film at the outer tube surface must always be less than the saturation temperature in order that any heat may flow through the condensate film. Even so, the temperature of the condensate in the hot well is often close to the saturation value, and the procedure for heating the condensate draining down through the tube bundle is to guide some of the exhaust vapor toward the lower part of the condenser, to provide what Silver [1963–1964] has termed *regenerative action*. Thus, the degree of subcooling is very much dependent on tube-bundle design, as well as the design and location of steam lanes.

Subcooling tends to increase with reduced cooling-water temperatures and, in some cases, can be controlled by adjusting the amount of circulating water bypassing a cooling tower. However, some plant operators believe that subcooling can be reduced by reducing the cooling-water flow rate. Unfortunately, this has the same effect on heat transfer as fouling and, due to the lower tube velocities, may give rise to an actual increase in solids deposition on the tube surfaces, so that a lower water flow may actually induce fouling!

9.2.2 SHELL SIDE

High Air Concentration in Condenser Off-Gas

The presence of excessive air in the condenser off-gas can be determined by comparing the temperature of the vapor with the saturation temperature, as follows (see Appendix H of the ASME standard [1998]). Let:

$$W_M = \frac{\text{unit weight of vapor}}{\text{unit weight of the noncondensable gases}}$$

m_V = molecular weight of mixed vapor

m_G = molecular weight of noncondensable gases

P_V = total pressure

P_T = saturation pressure as a function of the temperature

of the mixed vapor

Then

$$W_M = \frac{m_V}{m_G} \cdot \frac{P_V}{P_T - P_V} \tag{9.5}$$

Some non-condensible gas is always present in any condenser, and allowance is made for it in the design. However, if the concentration has increased above normal values, this may be due to air inleakage at some point around the turbine and/or condenser, the air finding its way into the condenser tube bundles. Air leak detection tests can be used to locate the source of the leak(s) so as to eliminate them (see Chapter 12). The air concentration can also increase because the air-removal system is extracting the vapor/air mixture at a slower rate. To establish whether this may be a factor, the vapor flow rate should be checked. Section H.5 of the ASME standard [1998] also contains an equation to estimate the volumetric flow of dry air in the off-gas.

Note that an increase in air concentration, or air binding, within the condenser shell causes the shell-side thermal resistance to increase, just as fouling causes the water-side film resistance to increase. Thus, both appear to have the same effect, and it is sometimes difficult to tell which is the primary cause of the overall thermal resistance increase.

Increase in Condensate Dissolved Oxygen Concentration

An elevated dissolved oxygen level usually accompanies an increase in the amount of air present in the shell side of the condenser and can be remedied by eliminating air inleakage, and/or by improving the operation of the air-removal system. Subcooling can also increase dissolved oxygen levels.

Fouling of Outer Surfaces of Condenser Tubes

Fouling of outer surfaces is uncommon but can occur, especially in geothermal plants or as a result of faulty boiler water chemical treatment.

Increase in Heat to Be Removed from the Turbine Exhaust

Damage to the turbine blading can sometimes cause the exhaust enthalpy to be greater than expected. If the turbine governor is able to hold the load, the exhaust flow may also have increased. The expected exhaust flow rate may be calculated from a turbine heat balance or from the heat/mass balances contained in the thermal kit. Compare also the values of the exhaust flow (present and calculated) shown in Table 9.1.

9.2.3 TUBE LEAKS

Change in Condensate Salinity

An increase in condensate salinity will normally be due to leakage of *circulating water* into the shell side of the condenser. The leakage, in turn, may be due to corrosion of the tube walls or leaks between tubes and the tubesheet. If not corrected, excess salinity will contaminate the water supplied to the boiler or steam generator and cause further problems. Leak detection procedures, using helium or sulfur hexafluoride as the tracer gas, can locate the source of these leaks (see Chapter 12).

9.3 DIAGNOSTIC METHODS

Analysis of the problems associated with the condenser on a particular turbo-generator will have to take into account the idiosyncrasies of the system involved and any site-related factors. It is difficult to present a general method which will apply to all plants. However, there are some common techniques for identifying probable reason(s) for changes in the condition of a condenser:

- Determining the reference conditions
- Experience and intuition
- Narrowing the search

- Diagnostic flowcharting to trace the causes of deviations
- Use of a fuzzy inference adviser

These techniques are among the steps which might be taken to diagnose with a high probability the possible cause(s) of a change in condenser performance and the action, if any, to be taken to rectify it. Clearly, while fouling is an ever-present problem and cleaning the condenser is often the means by which performance can be improved, it is not automatically the cause and must be considered among other possible causes, if the most suitable and effective maintenance decision is to be made. Note, too, that the techniques, as described in the following subsections, apply mainly to single-compartment condensers. The unique issues related to multicompartment condensers will be discussed separately, in Section 9.3.6.

9.3.1 Reference Conditions

The first step is to determine the set of reference conditions against which current conditions are to be compared, as a clue to finding where the problem lies. The operator may look to the design data as a reference, after making quite sure that the unit has not been modified significantly since its installation. At one plant, for example, a replacement low-pressure stage rotor was found to be less efficient than the original, causing the heat removal load to be increased. Significant information may also be obtained by calculating the set of conditions to be expected from a clean condenser operating under current conditions of generated power, cooling-water inlet temperature, and flow, using the methods outlined in Chapters 4 and 5.

9.3.2 Experience and Intuition

Familiarity with a particular unit will allow experience and intuition to be brought to bear on the data. However, the operator should keep in mind that new and unexpected factors may be contributing to the present problem and should thus not jump to conclusions based only on experience.

9.3.3 Narrowing the Search

In a given situation, some of the factors listed in Section 9.2 can probably be eliminated very quickly. First of all, the observed behavior can often be readily assigned to either the water side or the shell side of the system. And with adequate instrumentation in place, a review of the instrument readings, combined with proceeding through a short checklist, can indicate that certain of the factors are, in fact, not playing any part in the observed change in behavior. For instance, the waterbox levels can readily be checked, as can the waterbox differential pressures. Similarly, if the dissolved oxygen reading is within tolerances, it is improbable that the shell side has developed a problem.

Other clues can be found in distinguishing between gradual and sudden changes. The set of performance data taken at any point in time provides only a snapshot of the plant at that moment. However, a sound diagnosis must take account of how rapidly an apparent problem may have developed. For instance, a known and slow increase in fouling resistance will point toward fouling as the limit on generation capacity; while a sudden increase in air concentration in the off-gas will tend to point toward air inleakage as the problem.

Finally, requiring that some additional tests be conducted off-line, e.g., heat transfer tests and/or deposit analysis, can also augment the information on which a plan of action is to be based, and thereby further substantiate or eliminate possible causes. Thus, eliminating as many factors as possible can lead to less time being needed to identify the fundamental cause of a problem.

9.3.4 Diagnostic Logic Diagram

In some plants, a diagnostic logic diagram has been prepared to assist the engineer and/or operator in arriving at a proper conclusion. A typical diagram is shown in Figure 9.3, but the details have to be adjusted to match the available instrumentation and the configuration of the condenser subsystem and its auxiliaries. This diagram applies only to direct physical measurements and does not include any parameters derived from models or calculations. For instance, from the diagram, fouling can be supported as a possibility if the back pressure and terminal temperature difference are both high and the dissolved oxygen concentration is low. However, an increase in the water differential pressure across the condenser can also indicate fouling, as can a rise in the value of c calculated from Equation (9.4). Further evidence of fouling may be obtained from the calculated value of the fouling resistance or from an increased value of the pump motor current.

A reduced water flow rate may be due to fouling of the tube bundle or of the tubesheet, to insufficient pumps in operation, to impeller wear, or, perhaps, to improperly positioned valves. As already mentioned, the indicator lights on the control panel can be misleading, sometimes indicating that a pump is on when, in fact, it is not rotating. Water flow problems may also be confirmed by a high pump discharge pressure; and further confirmation may be obtained from the water flow rate calculated in accordance with Equation (4.24).

Of course, an increase in the concentration of non-condensible gas can also appear to produce a result similar to fouling. To check this possibility, reference should be made to the concentration of non-condensibles in the off-gas, the off-gas vapor flow, and the dissolved oxygen concentration.

Unfortunately, the logic diagram for the whole condenser subsystem is usually much more complex than shown in Figure 9.3. Further, logic diagrams with yes/no decision branches allow the diagnosis to descend to only one malfunction or fault at a time. If multiple faults exist simultaneously, these diagrams become much more complicated. To simplify the task, it may be better to create a logic diagram for each subproblem or malfunction, later linking all the diagrams together by adding connections to appropriate decision branches.

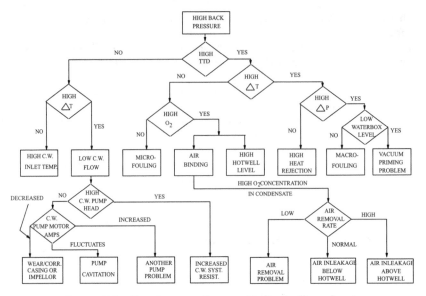

FIGURE 9.3. Condenser diagnostic logic flowchart.

Another approach is the use of decision tree analysis [Raiffa 1970], with probabilities assigned to the decision branches rather than mere yes/no decisions. In the final analysis, the solution to a problem is initially a probability, not a certainty, especially when there might be more than one malfunction present. After all, air inleakage can occur in the presence of fouling. The rate at which parameters have been changing over time can also be an important indicator of whether the change is recent. If a plot of the fouling resistance shows no signs of a sharp increase, then the rise in condenser back pressure may be due to air inleakage.

9.3.5 Fuzzy Inference Adviser

Katragadda [1997] outlined a *fuzzy inference adviser* which had been developed under the auspices of the Electric Power Research Institute (EPRI). Based on the principles of fuzzy logic, it uses a specially designed software program loaded into a personal computer, having the ability to receive the data set from the operating unit on-line. The software database is customized to reflect the set of logic statements appropriate to a given unit: the set of sigmoid (membership) functions must also be customized. The program becomes much more complicated when the condenser has more than one compartment.

If used only for units running under base-load conditions, it is possible that such systems can be calibrated once and will then provide satisfactory advice. However, for cycling units, the values of the set of reference parameters will vary with load, and such systems have to be designed to receive them. This also means that the significance of deviations will be load-dependent, so that the membership functions may need to be adjusted with respect to load. Thus, a fuzzy inference adviser program is not a "black box" which can quickly be adapted to match the behavior in a given unit environment. It is possible that such devices will provide useful assistance to future unit operators, but the knowledge engineering required to apply such techniques to a given plant costs so much that these systems are not currently cost-effective.

9.3.6 Multicompartment Condensers

The analytical task is made more complicated when the condenser has more than one compartment, especially when the compartments operate at different back pressures. For instance, it is possible that a high back pressure will be of concern only in the compartment from which the cooling water is discharged while conditions in the compartment connected directly to the cooling-water pumps remain well within range. In such a case, because of the differences in the mean compartmental water temperatures, fouling can be more severe in the compartment which is operating at the higher back pressure.

Condensers and their piping systems are not usually designed so that dissolved oxygen concentrations can be measured independently in the hot wells, since the condensate stream often cascades from one compartment to another internally before being collected in a common hot well. Furthermore, the air-removal system is often designed to serve all compartments so that the concentration of noncondensible gas in the vapor withdrawn from a particular compartment cannot be determined with certainty. This makes it difficult to associate a suspected air ingress with one particular compartment.

In those condensers where tube bundles extend continuously through all compartments, it is not physically possible to measure the mean water temperature or the local values at the compartmental interface(s). However, even with condenser compartments arranged in accordance with Figures 3.4 and 3.5, seldom are sensors furnished to measure the intercompartmental cooling-water temperatures. The models outlined in Chapters 4 and 5 do permit most of the gaps in the information to be completed, not only for a clean but also for a fouled condenser, so that instrument omissions can often be overcome. Parameters estimated in this way provide more complete information to the operator and engineer and permit better judgments to be made about the true state of the condenser.

Chapter 10

FOULING, CORROSION, AND WATER CONTAMINATION

This chapter covers some of the chemical and environmental issues surrounding condensers: fouling and corrosion of tubes, and contamination of cooling water. A tendency toward tube fouling is a perpetual problem with condensers. Corrosion is also a concern; it is sometimes initiated by fouling, as in the case of underdeposit corrosion. Maintaining equipment performance, availability, and life expectancy depends on a complicated interaction between water chemistry, tube metallurgy, and the fouling mechanisms involved, together with condenser-system kinetics and thermodynamics.

Because similar issues also affect heat exchanger tubes and heat exchanger service water, the problems associated with heat exchangers and condensers will often be compared and contrasted within the chapter. Schwarz [1988] outlined how steam surface condensers and service-water heat exchangers have different susceptibilities to fouling and corrosion. There are a number of equipment characteristics which can promote corrosion and/or fouling. In these respects, heat exchangers differ from steam surface condensers in at least the following ways:

- Heat exchangers have a lower tube heat transfer coefficient.
- They have smaller tube diameters.
- They are more likely to have multiple tube passes.
- Their tube-side velocities are variable and generally lower.
- Heat exchangers employ a variety of fluids on the shell side.
- Copper alloys are used extensively in their fabrication.
- Operate at higher temperature levels.
- Operate at higher fluid temperature differences.
- Operate at lower Shell-side velocities.
- Operate at higher pressure levels.

Fouling and corrosion are affected by equipment design, and they will also vary with the chemistry of the cooling or service water and according to whether the water is drawn from a river or the source is a closed-loop water circulation system, which will usually include cooling towers. Many fouling processes are sensitive not only to temperature but also to velocity. Thus, surface condensers usually operate with a design water velocity of 7 ft/s, sufficient to prevent buildup of fouling, except when pumps are switched off. However, service-water heat exchangers are frequently provided with a temperature-controlled valve which automatically adjusts the flow, and these generally operate at much lower velocities, especially during the winter months. Unfortunately, lower velocities can allow particle sedimentation, and fouling and/or corrosion often passes unnoticed until the water temperature rises in the summer months.

10.1 COMMON CAUSES OF FOULING AND CORROSION

10.1.1 Principal Modes of Fouling

Fouling of condenser and heat exchanger tubes can take several forms; the most important are:

- *Sedimentary fouling (silting)*. Sediment (particulate matter) entrained with the water can form deposits when the water velocity falls to the point where the sediment is no longer suspended. To avoid this, water velocities should be maintained as close as possible to their design values.
- *Salt deposition (crystallization)*. Inorganic salts of various kinds can be deposited due to changes in their concentrations, which are affected by the varying temperature of the water as it passes through the condenser tubes. Salt deposits typically consist of at least calcites, sulfites, silicates, or metal oxides of various valences, all of which lower the tube heat transfer coefficient and many of which are very difficult to remove once they have been deposited. In heat exchangers, fouling can also take the form of scaling, which is caused when the cooling water becomes saturated or supersaturated with the scale-forming ions. The initialization of crystallization (nucleation) is sensitive to tube-side velocity, as is also the thickness of the scale.
- *Microbiological fouling*. Microbiological fouling can occur when organisms (such as bacteria) that are rich in metal compounds, e.g., those of manganese, attach themselves to tube surfaces and then produce an impermeable layer. Not only is the heat transfer affected, but many such deposits can also cause underdeposit corrosion, even of stainless steels. This, of course, has a direct effect on tube life. Deposits of this nature are, again, difficult to remove. Microbiological fouling is also sensitive to both temperature and velocity, often tending to increase in thickness as the velocity falls. Because of generally lower water temperatures during the winter months, chlorination used to control biofouling is frequently reduced. Unfortunately, while this decision is normally based on the performance of the condenser, chlorination reduction may lead to an increased rate of fouling of heat exchangers.
- *Macrobiological fouling*. Macrobiological fouling can occur when mollusks, such as zebra mussels, or similar aquatic organisms pass through the screens and attach themselves either to the tubesheet or to the inside surfaces of tubes. These creatures tend to reduce water velocity in the blocked tubes (and so heat transfer), making them more susceptible to fouling of other types. Macrofouling from large organisms can occur in

both surface condensers and heat exchangers, but because the latter have generally smaller tubes, they are more susceptible to macrobiological fouling.

The rate at which each of these fouling mechanisms develops over time is very site-specific and also depends on chemical treatment methods. Thus, a knowledge of the relevant fouling factor versus time is necessary in order to develop an economic profile of fouling losses, and the losses must be taken into account when developing the cleaning schedule for a condenser over a period of, say, one year.

10.1.2 Some Causes of Corrosion

Corrosion can occur because the tube material is unsuitable for resisting chemical attack. Pitting corrosion of condenser tubes often occurs when the tube alloy selected was not appropriate for the type of cooling water circulating on the particular site. In some cases, changes in cooling-water chemistry and/or impurities will increase pitting; for example, for plants located on tidal estuaries, the salinity of the water can vary with the tides. Oxygen depletion in water left standing in equipment during plant shutdowns can also cause pitting corrosion, even of austenitic stainless steels; while, during plant outages, copper alloys can be susceptible to pitting from those biological activities which produce hydrogen sulfide.

Pitting is also one of two principal forms of heat exchanger tube corrosion, the other being crevice corrosion. Crevices can be created either by tube blockage or by deposits which are the cause of local oxygen depletions. If deposits are allowed to remain, chlorides will destroy any protective oxide film on the tube wall and concentrate within the deposit. Crevice corrosion in heat exchangers can certainly be reduced by keeping the tubes clean.

10.2 THE MONITORING OF FOULING AND CORROSION

It was shown in Section 5.7 that fouling can be monitored by calculating the effective fouling resistance of the tube bundles, using an interactive model of the condenser and the low-pressure stage of the turbogenerator. However, fouling can also be monitored by using specially designed equipment which tracks not just the progressive changes in heat transfer, but also the chemical and corrosion effects within the cooling-water system. This instrumentation can also be arranged to transmit the information to a remote site so that the monitoring of several units or even plants can be centralized. In nuclear plants, this apparatus can also be used to monitor service-water systems.

Cooling-water systems are often of the closed-circulation type using cooling towers for heat removal. Others are of the once-through type, in which a source of cool water from a lake, a river, or even the ocean is pumped through the condenser or heat exchanger and immediately discharged to the environment. Clearly, the chemistry of the cooling water presented to the system will vary in each case, affecting its impact on corrosion and fouling together with the water treatment method(s) selected.

Service-water systems in nuclear plants are often used to cool equipment only in emergency situations and, for this reason, are especially difficult to monitor and maintain. Recognizing this, the NRC issued Generic Letter 89-13 [Nuclear Regulatory Commission 1989], requiring all holders of nuclear operating licenses to regularly evaluate the condition of the emergency service-water systems in their plants, so as to ensure that they can remove the heat loads which will result from a postulated event. It was in response to this letter that plants began to study the impact of the service water on tube fouling and corrosion, by passing a stream of the cooling water through a monitoring system arranged to operate in parallel with the main equipment.

Some monitors have been built just to study fouling. In these units, a model condenser is used to track the changes in heat transfer coefficient (under condensing conditions) of one or more tubes installed in the monitor. The tubes receive a continuous sample of the water supplied to the main condenser or heat exchanger. Since the monitor operates in parallel with the main equipment, tubes can be removed for study and/or cleaning at any time, without any effect on the operation of the turbogenerator unit itself. From the acquired data it is possible to plot the tube fouling resistance versus time and compare one fouling model with another for different times of year and/or for different water treatment methods. On several occasions, these monitors have even been used just to provide an impartial referee between the results obtained from the different and competing water treatment methods offered by a variety of vendors.

Other units have been constructed to simulate the full length of the tubes installed in the main unit. One model used a tube in a model condenser which was 40 ft in length. Another monitor was designed with five tubes mounted in one shorter model condenser, each tube being 6 ft long. These tubes were arranged so that the water flowed through them in series, providing another way of closely simulating the behavior of the continuous 30-ft tubes installed in the main condenser.

Where corrosion is of major concern, the model condenser is often used in parallel with other sample streams, each containing corrosion coupons which have been immersed in a flowing stream of water. This arrangement allows a variety of water conditions in the main equipment to be simulated, so that the effects of both chemistry and water velocity can be studied.

Monitors containing model condensers have also been built for laboratory use to measure the heat transfer coefficients of new and used tubes, to compare the effects of different tube cleaning procedures, to establish the effect of lining or coating tubes, and to differentiate between the effects of fouling on both the insides and outsides of tubes.

10.2.1 A Monitor to Reduce the Cost of Chemical Treatment

Schwarz et al. [1992] noted that the annual cost to a plant for chemicals and chemical water treatment services can often rise to between $700,000 and $1 million. For this reason, one nuclear plant embarked on a chemical treatment program to establish the optimum procedure for minimizing the fouling of surface condensers and service-water heat exchangers, as well as maintaining the carbon steel and 90-10 Cu-Ni tube corrosion rates within targets. However, vendor water treatment programs vary in their effectiveness. Therefore it was felt that data from a model cooling-water system would be needed to provide impartial evidence to evaluate the effectiveness of different programs.

The configuration of the model cooling system installed at the river-water pump house for this site is shown in Figure 10.1. It was a closed system complete with its own cooling tower and recirculation tank. Included was a model condenser, complete with the ability to measure fouling or deposition rate, and used both to simulate a condenser and to provide a model of a service-water heat exchanger. This installation also had several parallel test legs, which were used to study the effects of different velocities and tube materials, using both flat and crevice corrosion coupons, together with a linear polarization corrosion rate (LPR) meter.

At the end of the test of a given vendor water treatment program, the tubes were removed from the model service-water heat exchanger and evaluated for deposit weight and density, deposit morphology, and evidence of pitting. Corrosion was evaluated using both the coupons and the LPR meter, but, due to the short run length of 18 days, only the LPR was used to evaluate the general corrosion of carbon steel and 90-10 Cu-Ni. It was found that, while the corrosion rates determined by the coupons agreed well with those determined using the LPR, the coupon values for Cu-Ni were generally lower than the equivalent LPR value.

10.2.2 A Monitor to Study Corrosion and Suspended Solids

Davenport et al. [1992] reported on a program to study the treatment of the service water in another nuclear plant. The principal objective was to select the type of chemical treatment program and to optimize the dosage rates for individual chemicals. Thus, the design of the monitoring unit focused on measuring the impact of chemical addition on the suspended solids and on the fouling and corrosion rates. In particular, the unit was intended to study the impact of various polymers on suspended solids accumulation, as well as the impact of oxidizing biocides on carbon steel and copper alloy corrosion. By using a sidestream monitor, these factors could be evaluated through pilot plant testing rather than by full-scale chemical treatment of the entire service-water system.

The configuration of the monitor supplied to this plant generally followed Figure 10.2, which shows all three of the "legs" provided with this installation. Leg A was the control leg and had no chemical injection; leg B was treated only with a dispersant for suspended solids control; while leg C was treated with the dispersant as well as with an oxidizing biocide, such as sodium hypochlorite. Each leg contained a differential-pressure-type fouling monitor, a set of corrosion coupons, and a linear polarization rate (LPR) monitor, each provided with

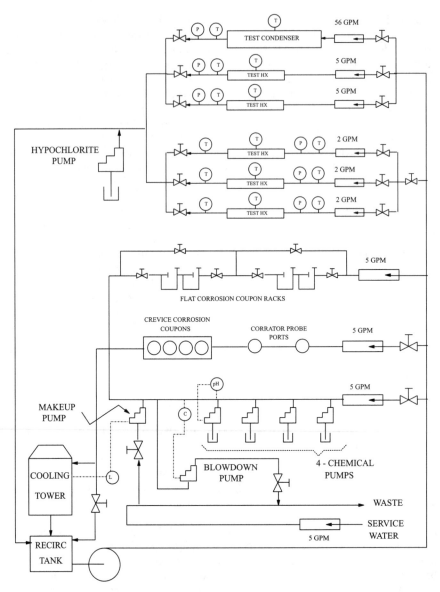

FIGURE 10.1. Model cooling system.

closely spaced three-electrode probes. The fouling monitor included in each leg consisted of a 48-in length of tube, made from the same material as the tubes installed in the main heat exchanger. The flow rate through each tube was carefully controlled, so that increases in differential pressure could be attributed entirely to the effects of fouling.

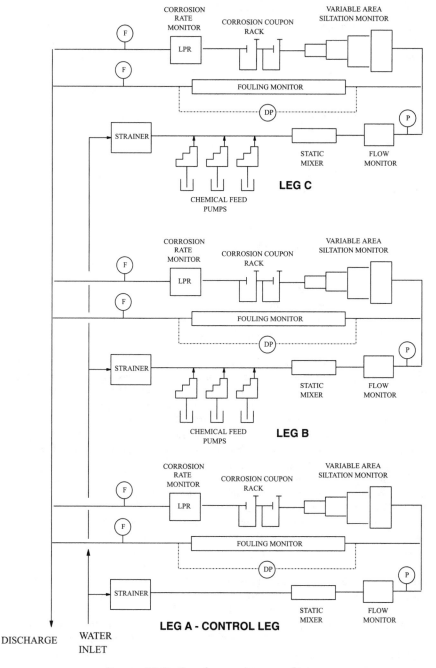

FIGURE 10.2. Service-water monitor.

The variable-area siltation monitor shown in Figure 10.2 consisted of several sections of pipe with decreasing cross-sectional areas arranged in series, the velocity of the water increasing as it flowed from one section to another. The type of dispersant determined the location of the section in which solids were deposited, as well as the velocity at which it occurred. Data from this unit showed that:

- An increase in water temperature strongly corresponded to an increase in the carbon steel corrosion rate.
- The difference between the corrosion rate in the control leg (leg A) and that in the dispersant-treated leg (B) was not statistically significant.
- Leg C, which was treated with hypochlorite, showed a significant increase in corrosion rate.

10.2.3 A Model Service-Water and Circulating-Water System

The service-water system in still another nuclear plant received water from a radial well system, which reached well under the Mississippi River and so tended to have fewer suspended solids, any associated impingement problems also being reduced [Antoine 1991]. However, the water was high in iron and manganese, and also contained bacteria (*Gallionella*) which thrived on iron. Several solutions to the water treatment problems had been evaluated, but the lack of an objective basis for comparison made it difficult to determine the best option. To remove the uncertainties it was decided to install a physical model of both the plant service-water and circulating-water systems on-site and provide the model with sufficient and accurate instrumentation, the data from which would allow a proper comparison between different approaches. The monitor for this plant, shown in Figure 10.3, was designed to simulate not only a plant service-water (PSW) heat exchanger but also a portion of the main steam surface condenser.

The PSW model consisted of a single-tube heat exchanger in the form of a model condenser, intended to simulate the component cooling-water exchanger (CCWX), customarily the first exchanger to show signs of fouling and the one to require the most frequent cleaning. The tube was made from 304 stainless steel, as in the actual CCWX; while the water velocity through this unit was designed to be 2 ft/s, in order to simulate extremely low flow conditions.

The circulating-water system was simulated using a model condenser containing four tubes and operating under condensing conditions. The water to the model condenser was arranged to simulate the conditions at the inlet to the high-pressure condenser, where, historically, the high temperatures at this point in the cycle had caused tenacious deposits of iron and manganese to accumulate. The unit was designed to operate at a tube-side water velocity of 6.6 ft/s, and the material of these tubes was also 304 stainless steel, the same as in the main condenser. As shown in Figure 10.3, the model condenser was installed in its own closed circuit complete with cooling tower.

A source of untreated water was connected to the skid and, after being treated

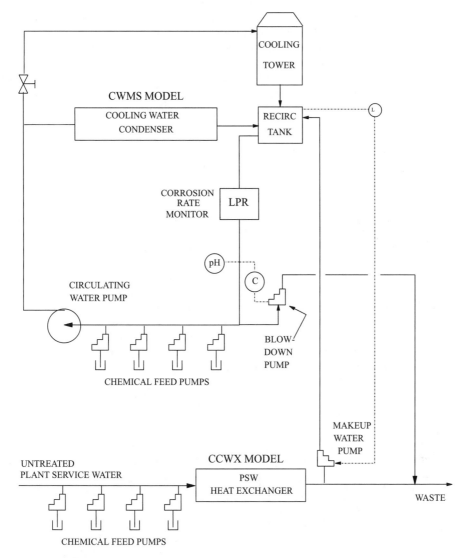

FIGURE 10.3. Model plant service-water (PSW) and circulating-water systems.

in accordance with vendor recommendations, entered the plant service-water heat exchanger (CCWX), on which the inlet and outlet water temperatures, shell temperature, and vacuum and corrosion rates were measured. The discharge from the CCWX also provided a source of makeup water to the circulating-water model subsystem (CWMS). In the CWMS, chemicals were added before the circulating-water pump, while a conductivity probe, a pH probe, and a corrosion rate meter were also included in this subsystem. Other instrumentation around the model condenser included sensors for measuring inlet and outlet water temperatures, shell temperature, and vacuum.

The success of this monitoring program lay in providing adequate and reliable data which allowed engineers and analysts to differentiate between effective and ineffective water treatment programs, as well as the ability to duplicate the conditions and types of foulant to be found in the actual plant.

10.2.4 A Mobile Cooling-Water Monitoring System

Another utility wished to obtain a customized mobile cooling tower water monitor, designed to be moved from plant to plant, so as to study the uniqueness of the water problems associated with each plant, whether fossil fuel–fired, combined-cycle, or nuclear. The configuration of this system is shown in Figure 10.4. Here the cooling tower, makeup, and blowdown water pumps were located on one portable skid; while the model condenser or heat exchanger, corrosion coupons, LPR corrosion meter, chemical injection pumps, and data acquisition system were all mounted on a second skid. The data acquired at the site was to be monitored remotely from the engineering headquarters, regardless of the location of the site at which the unit was currently operating. To do this, the data acquisition unit was equipped with a modem and an RS-232 link complete with communications protocol and the capability to transmit data files to a remote personal computer on requests from the central office.

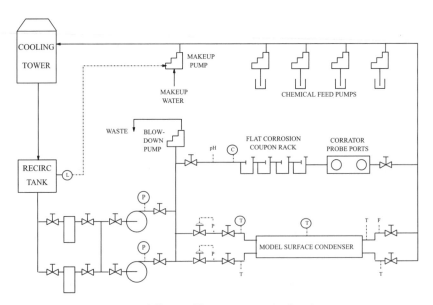

FIGURE 10.4. Mobile cooling-water monitoring system.

Several tubes of different materials could be installed in the model condenser, and the effective heat transfer coefficient of each individual tube could be calculated on-site from the acquired data. The procedure contained in the new ASME standard was to be used to obtain the reference heat transfer coefficient for each tube, but, to simplify the equipment, this could only be calculated by the personal computer located in the engineering center, and after a data file had been received from the site. Because both the water velocity and heat input rate to the model condenser could be adjusted on-site, it was also possible to study independently the effects of these two parameters on corrosion and heat transfer.

High-accuracy instrumentation was chosen in order to increase the confidence level in the data and calculated results. A number of automatic controls were included to ensure that (1) the water inventory in the closed system remained within tolerances, (2) the chemical injection rates could be adjusted and measured, (3) the pH of the circulating water could be controlled at preset values, and (4) the circulating-water temperature could also be controlled at a selected value. The mobility of the unit and its ability to be monitored from an engineering center placed the responsibility for the performance and maintenance of the unit and the integrity of the data on just a very few people. This approach allowed the overall costs of the experiments to be minimized, and their planning and execution to be managed by a dedicated group.

10.2.5   Portable Test Condenser

The portable test condenser was the first use in the power industry of a model condenser to monitor fouling and corrosion, the design of subsequent units being based on the experience gained from this original application. The configuration of the portable test condenser is shown in Figure 10.5. The model condenser was designed to accept only one 36-in-long sample tube, through which was passed a continuous stream of the cooling water whose impact on the tube was to be studied. To improve the accuracy of the monitor, a flow control valve was used to set and maintain a given flow rate, thus eliminating an important potential disturbance. The heat transfer coefficient was calculated in an onboard computer which formed an essential part of the monitor. The computer was programmed to calculate the effective heat transfer coefficient at regular intervals, using the measured flow rate together with the signals from accurately calibrated temperature-sensing devices. The source data and calculated result were then stored as a time-tagged set in the memory of the computer. By using a personal computer equipped with a modem, and coupling it to the RS-232C data link included with the onboard computer, the accumulated sets of data could be retrieved remotely on demand, and then processed.

10.2.6   Monitoring of Fouling and Corrosion—Summary

The combinations of water quality, equipment characteristics, metallurgy, and possible water treatment procedures are so broad that no one design of a water

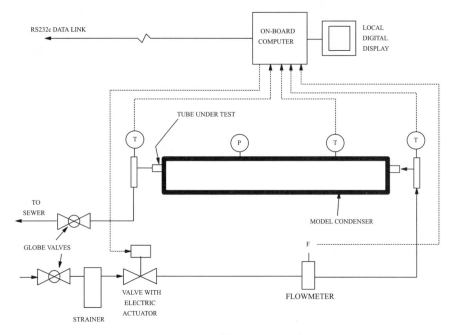

FIGURE 10.5. Portable test condenser.

monitoring system can meet the desire for "one size fits all." However, one constant feature of almost all these monitors has been the inclusion of model condensers of various sizes and capacities, the model condenser supplied with the portable test condenser being the most versatile form. These customized designs have been based on experience in the analysis of corrosion and fouling tendencies of different types of water, the impact of water chemistry on the plant equipment and its metallurgy, sound practice in the application of instrumentation and control, the design and programming of data acquisition systems, and the application of appropriate data communications technology where data is to be processed in another location.

10.3 GALVANIC CORROSION OF STAINLESS STEEL TUBES

Stainless steel tubes are liable to become corroded when immersed in any type of water. The corrosion is normally galvanic, in which an electromotive force is generated between two dissimilar metals, the cooling water acting as an electrolyte.

According to the theory of galvanic corrosion, metals and alloys can be arranged in a series in which magnesium, zinc, and aluminum all lie toward the *active* end while titanium, passive stainless steel, and copper alloys lie toward the *passive*, or noble, end. When a galvanic couple exists, the less noble partner acts as an anode and will tend to corrode. The relative surface areas of anode and cathode are important in determining the corrosion rate: if a high cathode area is associated with a small anode area, the corrosion rate at the anode is increased. Corrosion rates also rise with increased water salinity and/or temperature.

Note that stainless steel can be in either an active or a passive state. In air, stainless steel is passive. However, when immersed in water, stainless steel is normally active and less noble than copper alloys but slightly more noble than carbon steel. To bring stainless steel to a passive state requires the application of anodic protection.

10.3.1 The Potential for Galvanic Corrosion between Manganese and Stainless Steel

Condenser cooling water frequently contains various types of manganese compounds and even manganese in a biological form. Tombaugh [1989] indicated that manganese dioxide deposits can occur in condensers when the cooling water is drawn from rivers which also receive acid mine drainage. Pennsylvania and West Virginia are two states where this is frequently the case. Manganese dioxide is also found in cooling-water reservoirs which have been constructed over pine forests, and this source is common in the South. Inorganic manganese dioxide is a black solid and is feebly acidic. However, it can often be found in nature in association with compounds of iron, so that the color of deposits can range from black to red, depending on the amount of iron present.

Inorganic manganese dioxide can deposit on heat exchanger surfaces through the oxidation of manganese by exposure to dissolved oxygen or through overchlorination. A more organic form of manganese dioxide, however, can occur through a process known as biomineralization [Little 1998]. MnO_2 biodeposition can occur when certain types of microorganisms grow in the source of the cooling water, concentrate, and then convert soluble manganese to insoluble oxides, which are deposited on heat exchanger surfaces in the form of a black sludge. Biocides can exercise some level of control over this source of manganese dioxide deposition, but the biocide can itself encourage manganese dioxide formation.

Manganese dioxide is a corrosive substance for stainless steel whatever its origin and regardless of the nature of the cooling water. Not only do manganese dioxide deposits lower tube heat transfer coefficients, but they are also very difficult to remove: because they adhere so firmly to the surface of stainless steel tubes, they are seldom removed completely. Manganese dioxide deposits also encourage underdeposit corrosion, which can, however, be prevented or suppressed if the tubes are maintained in a clean condition.

Once manganese deposits form on the tube surface, they prevent oxygen from reaching the metal and thus from restoring the oxide film over it. Furthermore, ions such as chlorine, fluoride, sulfide, or permanganate migrate into the defect site and exacerbate the pitting problem. Because the area of the pit is much smaller than the surface area of the stainless steel, pitting can proceed rapidly once it begins.

Anderson [1990] reported on results of experiments with stainless steel tubes. The manganese deposits were removed using carbon steel mechanical scrapers. No transfer of carbon steel onto the stainless steel tube surfaces was observed, and this has been the common experience.

According to Sedricks [1979], manganese is detrimental to the pitting resistance of stainless steels in the presence of chlorides. It is also detrimental to the oxidation resistance (i.e., natural passivation) of these steels, $MnO \cdot Cr_2O_3$ being formed in preference to a protective Cr_2O_3 scale.

Szklarska-Smialowska [1986] reports that crevices often form around Mn-rich sulfide inclusions in the steel, which may occur even in Mn-poor steels. Because MnS can be detected in all pits and regardless of its concentration in steel, it can be concluded that pits form predominantly at Mn-rich inclusions. Note that manganese, besides being deposited, is also a constituent in many stainless steels themselves, in concentrations of up to 0.8%. Sulfur is also an inclusion in stainless steel, since not all can be removed during the steel manufacturing process.

Morris [1985] indicates that, even though oxidation of manganese occurs at a moderate rate, the reaction kinetics are temperature-dependent. The reaction is autocatalytic in nature; i.e., the reaction is catalyzed by any manganese compounds already present. Thus, although it may take a long time for deposits to appear, once they do, then the reaction will proceed rapidly to a heavy deposit, which is tightly absorbed onto the surface. Morris concludes that the best method for avoiding pitting as well as a deterioration of the heat transfer coefficient is to keep the tubes clean, especially from manganese deposits.

Manganese deposits can be removed using mechanical cleaners [Tombaugh 1989, Anderson 1990] or chemically [Szklarska-Smialowska]. However, it is difficult to determine just when to remove them in order to avoid the onset of pitting. Considering the autocatalytic nature of the manganese dioxide deposition process, there is a possibility that monitoring the thermal resistance of the deposit will allow the point just prior to rapid buildup to be established. If tube cleaning is subsequently performed, an unacceptable accumulation of manganese dioxide on the tube surfaces could be avoided.

10.3.2   The Potential for Galvanic Corrosion between Carbon and Stainless Steels

Mechanical cleaners have long been used successfully for the removal of scaling and other deposits from stainless steel condenser tubes, cleaning them thoroughly and without harm to the tube. Normally, the blades of the cleaners have been made from carbon steel, although, occasionally, special orders for stainless steel blades have been filled. However, questions have sometimes been raised whether cleaners made from carbon steel are suitable for use with stainless steel tubes. Closer examination, supported by at least 30 years of experience in their use, indicates that tube corrosion through galvanic action between carbon and stainless steels has not resulted from the use of carbon steel cleaners.

An investigation of the possibly corrosive effect of carbon steel on stainless steel tube surfaces must also distinguish between two major cooling-water sources: (1) fresh water and (2) brackish or sea water having various saline concentrations. With fresh water, there is almost no galvanic corrosion with stainless steel *provided that the tube surfaces are clean*. Indeed, the 1987 *Metals Handbook* [ASM 1987] states that with fresh water, iron and steel tend to *protect* stainless steel when galvanically coupled. With saline water, because the stainless steel is more noble than carbon steel, only the latter should corrode.

The "Smearing" Hypothesis

It has long been known [ASM 1987, p. 1228] that iron occlusions, or embedded iron, can cause rust marks to appear on austenitic stainless steels (e.g., 304 or 316) when the surface is wetted. This kind of corrosion is especially obnoxious when stainless steel is used in the fabrication of food manufacturing equipment. It is normally considered that the embedded iron is acquired during the metal-working processes (e.g., during rolling), and it is subsequently removed during manufacturing in a pickling process stage.

At about the time martensitic stainless steels were being introduced to the condenser market, there was a suggestion that "smearing" of carbon steel on austenitic stainless steel tube surfaces could occur as a cleaner passed down the tube, and that this might be a cause for galvanic corrosion between the two types of steel, as had been the experience with embedded iron. According to the smearing hypothesis, corrosion was supposed to occur only at a temperature greater than 179°F. This further casts doubt on the validity of the smearing hypothesis, since the maximum back pressure allowed by turbine manufacturers is about 5.5 in-Hg, for which the saturation temperature is 137.4°F. The only way in which higher temperatures could be experienced on the outer surface of condenser tubes is if steam is dumped following a unit trip. However, this steam is desuperheated, and temperatures approaching 179°F would be experienced only for a very short time. It is interesting to note that an extensive literature search revealed no references at all to tube corrosion through smearing.

Field Experience with Carbon Steel Cleaners

Stainless steel tubes have been cleaned by carbon steel scrapers for over 30 years without any evidence of harm to the tube surfaces. This finding was confirmed in a report by Anderson [1990] on experiments on stainless steel tubes which had been cleaned with carbon steel mechanical scrapers. No transfer of carbon steel onto the stainless steel tube surfaces was observed, and this has been the common experience. Furthermore, Sedricks [1979, p. 74] states that, although carbon can be detrimental when precipitated as a carbide or cementite, it has no effect on pitting resistance when in solid solution (e.g., as in a tempered carbon steel).

Should any uncertainty remain, alternative cleaners are available with blades made from stainless steel. However, cleaning quality may suffer if the deposits should be found hard to remove. Finally, the *Metals Handbook* [ASM 1987] also indicates that titanium and its alloys are strongly passive (noble) and, should any carbon steel be present, the latter would form an anode and only it would corrode, not the titanium.

10.4 POTENTIAL COPPER CONTAMINATION OF PRISTINE WATERS

The Environmental Protection Agency (EPA) has determined that the presence of certain levels of copper in pristine waters is detrimental to the health of aquatic life. This section examines the background behind these findings, outlines the monitoring techniques which will warn the power plant when the copper discharges from condenser cooling-water effluent are becoming too high, and suggests some approaches for reducing the concentration of copper compounds in condenser cooling-water effluent.

10.4.1 Effect of Copper on Aquatic Life

Copper is required in trace amounts for the growth and functioning of microorganisms, since it is a cofactor for numerous enzymes. Also, proteins containing copper are important electron carriers. Copper is also a useful trace element for human health, and existing standards allow a copper concentration in drinking water of up to 1300 ppb (parts per billion). However, the limit also varies by state; for example, Arizona allows a maximum of only 1000 ppb. Most of the copper in drinking water originates in domestic plumbing systems, and copper can be tasted at concentrations greater than 1000 ppb.

Unfortunately, elevated concentrations of copper can exert a toxic and lethal effect [Trevors 1990]. Research over a long period has shown that quite small concentrations of copper in fresh-water streams and rivers and in seawater can kill some forms of aquatic life and certainly reduce their rate of growth. One often cited study details the effect of copper corrosion products downstream from the Chalk Point power plant on the Patuxent River [Roosenberg 1969]. Lewis and Whitfield [1974] have also compiled an exhaustive review of the literature on the biological importance of copper in the sea. They give copper values for the North Atlantic which range from 2.98 to 12 ppb, while that for the English Channel ranges from 0.13 to 190 ppb. However, in oysters, the copper concentration ranges from 27 to 500 ppm (parts per million), which is very much higher than that of the water in which they live. This is not unusual, the literature indicating that many organisms concentrate copper and other heavy metals in the course of their metabolism.

Of course, the negative reaction of aquatic organisms to copper can be inferred from the almost universal use of copper-based antifouling paints for boats and marine equipment.

10.4.2 Sources of Copper Contamination

Contamination from Power-Plant Equipment

Although many power-plant condensers today are provided with tubes made from stainless steel, titanium, AL-6XN, and other noncopper alloys, some still operate with Admiralty tubes, 90-10 Cu-Ni, and similar alloys. Many plants also use copper alloy pumps and piping in the cooling-water system.

It is often suspected that an increase in the copper concentration in condenser cooling-water effluent is caused by contact with the exposed surface of copper-based alloy tubing. This is sometimes the case, and there are various remedies, which will be discussed later. Other studies have shown that there are instances of faulty plant or equipment design, together with some other factors which can cause copper contamination of condenser effluent. However, while some ecologists have used such data to condemn the use of copper tubing in steam-electric plants as always having a toxic effect on the nearby ecosystem, this is not necessarily the case.

Contamination in Fresh Water

The cooling-water source itself may contain copper compounds, e.g., with waters exposed to runoff from copper mines, as often occurs in Arizona. Durum and Haffty [1963] determined that the natural sources of copper entering the sea from the large rivers of North America averaged about 5.3 ppb. The copper content varies with stream depth, and other studies have shown that ionic copper can vary from 1 to 1.5 ppb and the total soluble copper from 10 to 15 ppb.

Contamination in Seawater

Copper exists in seawater in several forms: ionic, particulate, inorganic complexed, and organic complexed. Compton and Corcoran [1974] studied the effects of copper in a saltwater environment, and they quote copper values for the North Atlantic ranging from 2.98 to 12 ppb. In their study of six power plants located on the west coast of Florida, they found little distinction between the copper concentration in the Crystal River and that in the, presumably, polluted waters of Tampa Bay. The conclusion of their study was that properly operating steam-electric power plants do not cause damage to the nearby ecosystem by contamination with copper from their cooling systems. Other factors, especially silt or seasonal water temperatures, also play a part. For instance, in one case the heat in condenser cooling-water effluent was found to have seasonally killed young oysters; but during cool weather a profuse growth of oysters developed in the same general area.

Contamination from Closed Systems with Cooling Towers

Compton and Corcoran [1976] also conducted a study of closed cooling-water systems in which heat was removed by cooling towers. They found that the very nature of the closed cooling-water system, together with the local scarcity of water and other related factors, tended to preclude contamination of any local natural bodies of water. In most cases, there was no discharge, and where there was, the quantity of water was negligibly small.

In a cooling tower there are three sources of water loss: evaporation, drift or windage loss of small droplets of liquid water, and blowdown. The amount of blowdown is controlled to avoid a buildup of dissolved solids sufficient to cause scaling. The cycles or factors of concentration may vary from 2 to 20, for the most part determined by the calcium bicarbonate scaling tendencies, but other scaling considerations may also come into play.

Compton and Corcoran analyzed a total of 31 samples. Ten *makeup waters* had copper concentrations higher than 15 ppb, five were higher than 25 ppb, one was at 46 ppb, and one, 92 ppb. The mean was 13.5 ppb. Only two river waters were below 3 ppb. High concentrations over 1000 ppb were found in some recirculating waters, but not in most settling pond waters or in the effluent discharged to streams or other natural bodies of water. In no more than two cases of plants equipped with cooling towers was there a discharge with a copper concentration greater than the makeup water. Based on their analytical work and observations, they concluded that the subject plants were not a threat to the aquatic biota.

Some environmentalists have questioned the use of settling ponds and the method for disposing of the fill dredged out of the ponds, indicating that they think the copper is in a soluble form and so will continue to be dispersed through the groundwater. However, it should be pointed out that the copper in the blowdown from cooling towers discharged into settling ponds is oxidized to a cupric oxide and is, thus, extremely *in*soluble. The high concentrations in many of the sludge samples from the basins of towers also indicate a precipitation reaction in the tower.

Sources of Contamination—Summary

Analyzing the cause of excess copper in condenser effluent should begin by surveying *all* possible sources specific to the site, in order that the action taken will be truly appropriate to the given situation.

10.4.3 Monitoring Copper Concentrations in Condenser Cooling-Water Effluent

Water quality standards are set not only at the federal level but also by each state. They are modified from time to time, and steps should be taken to become informed whenever updates are published. At the state level, standards are set by the individual state's department of environmental protection. At the federal

Table 10.1   Copper Concentration Limits

State	Fresh Waters	Saline Waters	Drinking Water
Alaska	Eq. (10.1) is used	2.9 µg/l	1000 mg/l
Arizona	0.05 mg/l	NA	1000 mg/l
California	0.01–0.02 mg/l	0.005 mg/l—6mo[1] 0.020 mg/l—daily max[2] 0.050 mg/l—inst. Max[3]	1000 mg/l
Florida	Eq. (10.1) is used	2.9 µg/l	1000 mg/l
Pennsylvania	0.1 mg/l	0.1 mg/l	1000 mg/l

California saline water limits:
   (1) Average of 0.005 mg/l over 6-months
   (2) Maximum average over any one day of 0.020 mg/l
   (3) Instantaneous value at any time to be no more than 0.050 mg/l

level, standards are set by the Office of Water Regulations and Standards, a division of the U.S. Environmental Protection Agency, the water quality standards [EPA 1988] being an example. The introduction to these standards states:

> Copper is an essential trace element required in plant and animal metabolism. Most natural waters contain copper at levels not known to have any human or aquatic toxicological effects. Concentrations of copper result from metal plating, mining, pesticide production, and electrical products industries. The toxicity of copper to aquatic life is enhanced by lower alkaline conditions and increased water hardness. Copper in sufficiently high concentrations does impart undesirable taste to water. [EPA 1988]

Some typical limiting values of copper for a few selected states are given in Table 10.1.

The quality criteria for copper in water are often specified as follows:

Freshwater aquatic organisms and their uses:

- Chronic:

$$Cu \le \exp[0.8545\ln(\text{hardness})] - 1.465 \quad \mu g/l \qquad (10.1)$$

- Acute:

$$Cu \le \exp[0.9422\ln(\text{hardness})] - 1.464 \quad \mu g/l \qquad (10.2)$$

which hardness is expressed as milligrams per liter of $CaCO_3$.

Saltwater aquatic organisms and their uses:

- Chronic:

$$Cu \leq 2.9 \mu g/l$$

Collecting Samples

Water samples are normally collected by the plant as "grab" samples, either routinely or on an as-needed basis. The EPA can also collect samples independently, but it is not unusual for the EPA to accept the analytical results obtained from the samples gathered by the plant. The working arrangement between the local EPA and the plant can vary from state to state, from plant to plant, or even from unit to unit within the same plant!

The Department of Environmental Protection for the state of Florida publishes standard operating procedures [Florida DEP 1992]. In their case, effective monitoring locations are specified in individual discharge permits issued to power plants and are based on the "best professional judgment" of the permit writer.

Sample Preparation and Analysis

Clearly, good practice requires that utility companies monitor potential copper contamination, preferably by taking samples at the same time and frequency (and, perhaps, same location) as the local environmental protection authorities. If monitoring is not already being carried out in the plant, a monitoring regime should be instituted, using one of the Standard procedures outlined below. Clearly, routine monitoring and archiving of copper data not only promotes staying within allowable limits and promptly correcting any excursions, but it also provides an independent check of any data submitted by the regulator. Thus, routine monitoring by the plant is a prudent preventive measure, helping to avoid potential penalties from any violations of effluent contamination limits.

The U.S. Environmental Protection Agency has published a manual for determining the metals contained in environmental samples [EPA 1994]. The methods the EPA identifies are as follows:

- *Standard 200.2.* Sample preparation procedure for spectrochemical determination of total recoverable elements. This is a detailed sample preparation method which applies to both solid and aqueous samples. It requires the drying of the sample and the addition of both nitric and hydrochloric acids and takes several hours.
- *Standard 200.7.* Determination of metals and trace elements in water and wastes by inductively coupled plasma-atomic emission spectrometry (ICP-AES). The instrument must be calibrated against known copper samples which have been carefully prepared. Standard 200.7 also

includes a typical profile of instrument emission intensity versus argon flow rate which is used to evaluate the emission results obtained when testing samples taken from the plant.

- *Standard 200.8.* Determination of trace elements in water and wastes by inductively coupled plasma–mass spectrometry. A collaborative study on this method was prepared by Longbottom [1994].
- *Standard 200.12.* Determination of trace elements by graphite furnace atomic absorption spectrometry with a stabilized temperature platform.
- *Standard 200.15.* Determination of metals and trace elements in water by ultrasonic nebulization inductively coupled plasma–atomic emission spectrometry (UNICP-AES).

A recent method being promulgated by the U.S. Environmental Protection Agency for the analysis of copper concentrations in *seawater* is contained in Method 1640 [EPA 1995]. It is based on the use of inductively coupled plasma–mass spectrometry similar to Method 200.8. However, Method 1640 places greater emphasis on avoiding sample contamination, starting from the gathering of the sample and extending through all stages of preparing the sample for analysis.

Clearly, the data obtained from a sample analysis should be carefully logged and plotted for management review.

10.4.4 Remedial Measures

Most states' departments of environmental protection are more interested in reducing copper contamination below the specified limits than in imposing penalties. Thus, if a plant's attention is drawn to a violation, the situation should be approached with the objective of obtaining a negotiated solution. For instance, in plants with multiple units, it is often quite possible for a different solution to be negotiated for each of the various units.

Remedial measures can take several forms, and the choice for implementation will be very dependent on the circumstances. Where contamination occurs only just after mechanical cleaning of the tubes, the remedial action takes the form of passivation, which should be implemented just after the cleaning event. In other plants, contamination from copper may be persistent and is often due to constant exposure of the tube surfaces to water having a low pH. In such cases, some form of chemical treatment should be applied to raise the pH. If this is not possible or is ineffective, the remedial action can take one of the following forms:

- Tube lining
- Tube coating
- Tube replacement

Copper Passivation Following Mechanical Cleaning

It is possible that, after mechanically cleaning a condenser with tubes made from copper-containing alloys, excessive copper concentrations, as defined in Table 10.1, will be found in the cooling-water effluent. If this is the case, a passivation procedure should be implemented, Most chemical companies have appropriate chemicals and procedures and we use one recommended by NALCO Chemical Corporation [1998] only as an example. The best inhibitor for copper-containing metals is tolyltriazole (e.g., NALCO's Super-Cool 1336) which allows an organic film to form on the metal surface. *Of course, before working with any chemical, the user should review the most current version of the material safety data sheet for that chemical.*

First, ensure that the passivation program is compatible with the current water treatment procedures. It is very important that microbiological control be maintained during the passivation process. It is recommended that a nonoxidizing biocide such as isothiazolin (e.g., NALCO 7330) be used during passivation. If an oxidizing biocide such as chlorine must be used, it should be added only after the addition and mixing with the nonoxidizing biocide. During the procedure, the pH should be maintained within the range 7 to 9.

It is important that the condenser system be isolated during the passivation procedure and that water containing the chemicals be continuously recirculated through the tubes. If a cooling tower is associated with the condenser, it is sufficient to secure the blowdown and run the circulating water pumps. If the condenser is of the once-through type, it should be isolated by closing the inlet and outlet water valves on the waterboxes, a separate pump being provided to recirculate the water.

The dosage, or concentration, of tolyltriazole should be maintained at a minimum of 23 ppm for a period of from 24 to 48 hours. If this is not possible, the following equation may be used (in the case of NALCO 1336) to determine the minimum dosage for any duration from 2 to 24 hours:

$$\text{Dosage (ppm) NALCO 1336} = -1.06 \times \text{duration (h)} + 48.9 \qquad (10.3)$$

The quantity of tolyltriazole (NALCO 1336) to be added can be calculated using the following example for a cooling system with a volume of 20,000 gallons:

$$\text{Vol (ml) NALCO 1336} = \frac{20,000 \text{ gal} \times 8.34 \text{ lb/gal} \times 23 \text{ ppm} \times 3785 \text{ ml/gal}}{9.91 \times 10^6 \text{ lb/gal NALCO 1336}}$$
$$= 1470 \text{ ml NALCO 1336}$$

This represents the *minimum* initial dosage. Subsequent dosages would depend on the measurement of the inhibitor residual.

The passivation process requires that the inhibitor adhere to the tube metal surface so that, as the process proceeds, the inhibitor becomes depleted. Thus, testing of the inhibitor residual should be performed periodically during the passivation process, since it is desired that the dosage be sufficient to ensure optimal passivation of the tube surfaces. Residual measurements are also necessary after the addition of an oxidant, which tends to consume inhibitors. The most common method for measuring residuals is the use of a spectrophotometer, but other methods are available, and the chemical supplier should be contacted for details.

Following passivation, the system should be returned to service and the normal chemical treatment method should be placed back in operation. If corrosion of the tubes is also a concern, even after passivation, then an appropriate corrosion-inhibiting program (e.g., dosing with ferrous sulfate) should be implemented.

Tube Lining

One way to prevent the cooling-water from coming into contact with a copper surface is to line the tube with a metal liner. Liners are often made from titanium, because of its relatively high thermal conductivity. Care must be taken that the outer surface of the liner is in intimate contact with the inner surface of the original tube after the liner has been inserted; mechanical rolling or hydraulic techniques are used to ensure this. Unfortunately, tests have shown that it is not uncommon to find the U coefficient of a lined tube to be as much as 30% lower than that of an unlined tube made from the same material. This penalty has been attributed to the residual space left between the liner and tube surfaces in which air can be captured.

Thus, before applying this remedy, the plant should know the heat transfer coefficient of the lined tube assembly, to make sure that the rate of vapor heat removal by the condenser will not affect the generation capacity of the unit after the tubes have been lined. To verify this, it is possible for heat transfer tests to be conducted in a laboratory on samples of condenser tubes, after removing them from the unit and fitting them with liners of various dimensions.

Tube Coating

Another way to line a tube is to coat it with an epoxy resin, a practice which has received the approval of several states' departments of environmental protection. Coating can be applied in the field by experienced contractors. An EPRI report [1997] reviews the practice of tube coating and its economic evaluation. Again, because the heat transfer coefficient of a coated tube can affect the generation capacity of the unit, laboratory heat transfer tests should be conducted on tube samples removed from the condenser and coated as they would be in use.

Estimates have shown that the life of a coating is about five years. For this reason, it is often cost-effective to coat the tubes twice. Of course, after the second coating the condenser may need to be retubed completely.

Tube Replacement

Tube replacement should be a last resort to solving a copper contamination problem. Unfortunately, should tube replacement be necessary, it is unlikely that the local authorities will permit existing copper alloy tubes to be replaced with new tubes having a similar metallurgical analysis. Among the preferred replacement materials are titanium and stainless steel, but other noncopper alloys which have an adequate thermal conductivity (e.g., AL-6XN) can also be used. Note that a copper credit may be available to the plant after the original copper-based tubes have been replaced.

10.4.5 Copper Contamination-Summary

The presence of excessive concentrations of copper in condenser effluent can be detrimental to the health of aquatic life. Environmental regulations and current enforcement policy therefore focus on maintaining safe copper levels, although these do vary from state to state. Clearly, regular monitoring by a plant of its effluent is a prudent precautionary measure, allowing excursions in the level of local copper contamination to be corrected promptly. However, should copper levels in the cooling-water effluent rise, there can be several sources of the contamination, and the situation should be carefully analyzed before any remedial action is taken.

Chapter 11

ON-SITE MECHANICAL CLEANING OF CONDENSER TUBES

11.1 INTRODUCTION

This chapter first reviews a number of techniques commonly used for on-site cleaning of condenser and heat exchanger tubes. The various metals available for condenser tubing are identified and their properties compared. The chapter continues with a discussion of the techniques and tools available for mechanically cleaning condenser tubes, followed by some guidelines on choosing a cleaning method. Toward the end of the chapter are some case studies which report the results of using mechanical cleaning methods on condensers in nuclear and other plants. The chapter concludes with a discussion of waterbox safety issues.

A number of on-site condenser tube cleaning techniques have been applied over the years. Data obtained from an independent survey of 100 plants (two per state) conducted several years ago [Hovland 1988] determined the percentage of plants using some of these methods. Table 11.1 summarizes these findings.

Cleaning tubes with metal cleaners is a method first introduced over 70 years ago. Its market share continues to grow because the original tube cleanliness is better restored by this than by other methods. Brushes are also widely used, but tend to remove less of the deposit than metal cleaners. Hovland (1978) charted the improvements in back-pressure deviation that a power plant obtained after

Table 11.1.   Plant Cleaning Preferences

Cleaning Method	Plant Preference
Metal cleaners	44%
Brushes	22%
High-pressure water	14%
Sponge rubber balls	14%
Chemical cleaning	5%
Other methods	1%

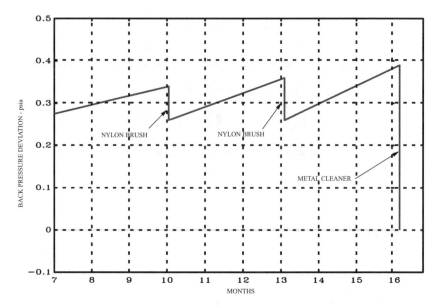

FIGURE 11.1. Back-pressure deviation improvement: nylon brush vs. metal cleaner.

two cleanings using brushes and one cleaning using spring-loaded metal cleaners (See Figure 11.1).

High-pressure water has often been selected for those condenser tubes which have very tenacious deposits. Unfortunately, the hydraulic lance has to be moved manually down the tube and at a rather slow rate, so that the time and expense of properly cleaning a condenser using this method tend to be high. Hovland (1978) also compared a power plant's improvements in back-pressure deviation obtained after separate cleanings using hydroblasting and spring-loaded metal cleaners (See Figure 11.2).

All three of these cleaning methods usually require that a unit be taken out of service, although it is sometimes possible to avoid a complete outage by reducing the load on a unit and then cleaning one waterbox at a time. Some methods allow cleaning while the unit is on-line. Sponge rubber balls are generally associated with the Taprogge on-line tube cleaning system. The sponge balls are often in the system for only part of a day [Kim 1993] and, rather than maintain absolutely clean tube surfaces, tend to merely limit the maximum degree of tube fouling. There is also some uncertainty concerning sponge ball distribution and, therefore, how many of the tubes actually become cleaned on-line. Mussalli et al. [1991] indicated that condenser waterboxes and inlet piping should be thoroughly modeled and the flow patterns studied with a view to improving the distribution of balls across the tubesheet. It is also not uncommon to find that numerous sponge balls have become stuck in condenser tubes, and these appear among the material removed during mechanical cleaning operations. For these

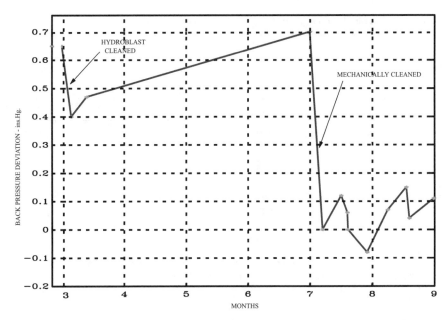

FIGURE 11.2. Back-pressure deviation improvement: hydroblast cleaning vs. mechanical cleaning.

reasons, the tubes of condensers equipped with these systems still have to be cleaned periodically off-line by other methods, especially if loss of generation capacity is of serious concern. It has also been found that, if sponge ball systems were not installed with the original equipment, then the cost-benefit ratio of adding them in the field tends to be high.

Several mildly acidic products are available for chemically cleaning condenser tubes and will remove more deposit than most other methods; but chemical cleaning is expensive, the procedure takes longer to complete than alternative methods, and the subsequent disposal of the chemicals—an environmental hazard—creates its own set of problems. It has also frequently been found that some residual material still needs to be removed by mechanical cleaning methods.

Finally, where fouling problems exist which are too severe to be handled by any of the foregoing methods, rotating tools with cutters, brushes, or scraping heads are available. However, they can gouge and otherwise damage tube walls if they are used aggressively or incorrectly.

11.2 TUBE MATERIALS AND THEIR FOULING AND CORROSION CHARACTERISTICS

Tube materials used for power-plant condensers and heat exchangers may be grouped into three categories:

Table 11.2. Tube Metal Properties, at 18 BWG (0.049 in)

Material	HEI Material Correction Factor	(1) Thermal Conductivity at 68°F, Btu/(h · ft · °F)	(2) Tensile Strength, psi × 10³	(3) Young's Modulus at 70°F, psi × 10⁶
Admiralty metal	1.00	64	50	16
Arsenical copper	1.02	112	40	17
Copper iron 194	1.03	150	60	17.5
Aluminum brass	0.99	58	60	16
Aluminum bronze	0.98	46	60	17
90-10 Cu-Ni	0.93	26	60	18
70-30 Cu-Ni	0.88	17	50	22
Cold Rolled carbon steel	0.93	27	45	29
SS type 304/316	0.75	8.6	70	28
Titanium	0.82	12.5	50	16

Note that experiments carried out by Purdue University show that the thermal conductivity of AL-6X, a form of stainless steel, should be 6.8 Btu/(h · ft · °F), the value adopted by Allegheny Ludlum in its literature.

- Copper-based alloys
- Stainless steels
- Titanium

All metals are able to transmit heat, and all metals foul, but their resistance to corrosion and their fouling rates vary, depending on the interaction between the chemical and/or biological characteristics of the cooling-water source. Thus, when selecting the metal to be used for the tubes in the condenser or heat exchanger, all of these factors need to be taken into account, in addition to tensile strength and modulus of elasticity.

Table 11.2 lists some of the physical properties used in tube metal selection [TEMA 1988, HEI 1995]. It is clear that, for a given cooling-water pump discharge pressure, the gage of the metal can be selected based on its tensile strength. This in turn affects the thermal resistance of the wall, as well as the cost of tubing, so that both the kind of metal selected and its gage (wall thickness) play a part in the detailed design of the tube bundles.

11.2.1 Copper Alloy Tubing

Tubes made from copper were probably among the first materials to be used to conduct cooling water through condensers and heat exchangers. However, native copper has a number of shortcomings, and several different copper alloys were developed to overcome certain specific problems which had become of concern to engineers over the years.

Among the first of the new alloys was 70-30 Cu-Zn, now known as *cartridge brass* because of its deep drawing characteristic. Unfortunately, it is not very resistant to seawater corrosion, and, around 1900, the British Royal Navy developed *admiralty brass*, which soon came into general use. It was found that, by adding 1% tin to the original 70-30 Cu-Zn alloy, its resistance to one of the most common causes of failure—dezincification—greatly increased. Then, during the 1930s, a U.S. brass company discovered that a small (0.04%) addition of *arsenic* further inhibited corrosion, and this became the subject of a patent. Other brass companies followed suit with additions of small amounts of *antimony* and *phosphorus*. Meanwhile, it should be noted that although admiralty brass became the most commonly used for fresh-water service, it is not considered acceptable for saltwater applications, due to its low erosion-corrosion resistance.

The further evolution of tube materials for condensers and heat exchangers resulted in the development of *aluminum brass* for saltwater service. Investigators took a copper-zinc alloy and added 2% aluminum, then raised the copper content to 77% and added a 0.04% dezincification inhibitor. These improvements caused aluminum brass to become the standard tube material for salt and brackish waters.

Copper-nickel alloys evolved when the U.S. Navy needed a tube material with better corrosion resistance to seawater. The Navy began to use tubes made from 70-30 copper-nickel alloys, and some power companies soon followed suit. Although no longer the case, the cost of nickel was initially a factor when these alloys were introduced. As a result, 90-10 Cu-Ni alloys were developed in order to reduce costs. The addition of up to 1.8% *iron* further increased the corrosion resistance of these alloys, which are reported to have 4 to 5 times the resistance to erosion-corrosion of the brass alloys.

One of the major drawbacks to using brass tubes in modern power-plant condensers is their susceptibility to *ammonia corrosion*. This occurs in areas of the condenser where ammonia can concentrate, for example, under the air-removal section or within the tube support plate annulus. However, while 70-30 Cu-Ni alloys are essentially immune to all forms of ammonia corrosion, there have been cases where 90-10 Cu-Ni alloys became damaged. Thus, while copper alloys have many advantages, they are also susceptible to several forms of corrosion, among them being:

- Crevice corrosion-pitting
- Dealloying-dezincification or denickelification
- Erosion-corrosion-either at the inlet end or further along the tube
- Ammonia corrosion-stress corrosion cracking

Crevice corrosion is the most destructive form of pitting of copper alloy tubes and is caused primarily by deposits. After years of service, pits can join together and form grooves or troughs in the metal walls. Samples of aluminum brass tubes which have been subjected to this form of corrosion often show very rough internal tube surfaces with very large-diameter shallow pitting and tube "washout." A typical cross section of a tube having an original wall thickness of 0.049 in (18 BWG) would show a variance in tube wall thickness of between 0.035 in and 0.049 in around the circumference of the pit. Such large areas of pitting con-

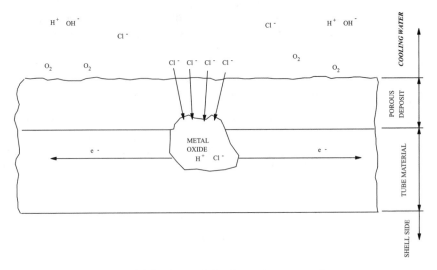

FIGURE 11.3. A model of crevice corrosion.

stitute crevice corrosion caused by deposits and can considerably shorten tube life.

Figure 11.3 is a model of crevice corrosion, showing the inside surface of the tube wall coated with a porous deposit after having been in service for a while. The water ionizes into hydrogen and hydroxyl radicals, some free oxygen is present, and the fluid often contains chlorine ions dispersed throughout. It is these chlorine and free oxygen ions which penetrate the porous coating and concentrate at the metal surface. The chlorine in the presence of free oxygen interacts with the base metal and causes oxidation and corrosion. If the concentrated chlorine and oxidation products could be flushed out (e.g., by cleaning the tube surfaces and removing the deposit), the extent of the corrosion could be controlled. However, being trapped beneath the surface of the porous deposit, the process continues and penetrates further into the metal wall. If allowed to continue, the corrosion will eventually penetrate through the tube wall.

Copper alloys *passivate* themselves by forming a protective cuprous oxide film on the tube surface. This film is nonporous and very thin and provides only a small resistance to heat transfer. If undisturbed, it will prevent the tube surface from pitting. Unfortunately, this film can become damaged during normal operations, e.g., scoured by sand or debris.

Unlike cuprous oxides, which form a film, cupric oxides grow continuously and can form a thick, porous layer on the tube inner surfaces. This oxide has a high resistance to heat transfer and disrupts the protective cuprous oxide film, helping the deposits to attack the base metal and resulting in the growth of any pits. These destructive copper oxide deposits must be cleaned out of all copper alloy tubes once or twice per year, allowing the protective cuprous oxide film to renew itself.

Copper alloy tubes are fundamentally sound basic materials for condensers and heat exchangers, and their corrosion can be controlled in cost-effective ways. The conditions at a particular site will determine the alloy selection. Combining this with the choice and frequency of an appropriate cleaning method can result in significant improvements in the heat transfer capability of a condenser or heat exchanger, increased unit efficiency, and extended tube life.

11.2.2 Stainless Steel Tubing

Although stainless steel has a lower thermal conductivity and higher resistance to heat transfer than copper alloys with the same wall thickness, it is generally more resistant to corrosion. Stainless steel tubes were first installed in surface condensers over 40 years ago, initially as peripheral tubes or in air-removal sections because of their superior corrosion resistance. The first condenser tubed entirely with T304 stainless steel was placed in service in a power plant on the Monongahela River in West Virginia, where acidic mine drainage had contaminated the river water. Still in service, the condenser on this unit has been cleaned on many occasions over the years, using metallic cleaners. An EPRI Report (1982) contained a survey showing that units accounting for more than 28% of U.S. generating capacity have condensers which are tubed with T304 or T316 stainless steel.

High unit availability depends on keeping condenser tube surfaces clean and free from deposits; otherwise crevice corrosion (or underdeposit corrosion) will result in premature tube failures, no matter what the nature of the cooling water. With stainless steel, crevice corrosion can be caused by deposits of iron, manganese, silt, or bacterial accumulations. The sensitivity of stainless steel to corrosion of this type was well known even prior to 1958. So great was the concern that many new plants had continuous ball cleaning systems installed to avoid the corrosion problem.

Chloride levels, the presence of deposits, and the water temperature all affect pitting and potential pitting rates. T304 and T316 stainless steels are especially subject to pitting when exposed to flowing water containing chlorides, as a result of localized oxygen depletion. The chlorine is transported through the porous deposit and destroys the protective film, while acidic conditions also form in the crevice as it develops. Because oxygen is not available to re-form the protective film, pitting will continue to progress and can only be controlled or suppressed by periodic tube cleaning.

The EPRI (1982) report investigating the causes of condenser tube leaks found that stainless steels have excellent resistance to corrosion when the condensers receive either fresh water or water recirculated through cooling towers, especially when the tubes are kept clean. Cooling tower water is heavily oxygenated, and its chemical condition can be easily controlled to reduce deposit formation and maintain an elevated pH level. Unfortunately, cooling tower chemistry can sometimes get out of control, and the rate of fouling deposition can increase during such periods.

Some more recent stainless steel alloys which have been developed by tube manufacturers include AL-6XN, Sea Cure, and AL 29-4C, all of which are considered to be superior to T304 and T316 stainless steels because of their increased resistance to crevice corrosion. They cost as much as, or even more than, the equivalent titanium tubes and are commonly used for salt- or brackish water applications.

Fouled tubes not only increase the costs of generating power, due to the higher heat rates which result, but tube failures due to corrosion can lead to large capital expense if the condensers have to be retubed. In addition to implementing a proper tube cleaning program, stagnant layup conditions should be avoided. During outage periods, the condenser should be drained and the tubes cleaned while the deposits are still wet, then flushed with fresh water. Alternatively, the cooling water should be circulated continuously during the outage so that it is not depleted of oxygen.

11.2.3 Titanium Tubes

It was in the early 1970s that the first power-plant surface condenser equipped with titanium tubes was placed in service. Titanium has since become the preferred material for condensers receiving salt or brackish cooling water, although it is also being used in plants receiving fresh water or even water from a cooling tower installation. Initially, titanium tubing was used essentially to replace existing copper alloy tubes. The first condenser to use titanium tubes in its original design did not go into service until 1978.

As with most new materials, a learning curve had to be experienced in order to obtain the optimum service and reliability, and this was especially the case with the titanium retubing applications. While the modulus of elasticity of titanium is the same as that of brass, the tube wall thickness was commonly only 58% of that used for brass tubes, in order to offset the lower thermal conductivity of titanium. However, in a condenser with tube-support-plate spacing designed for a thicker tube wall, there was the possibility that the high steam exhaust velocity would now induce tube vibration. No published information or approved procedures were available to determine the correct support spacing, and, in many cases, advice was not available from condenser manufacturers, many of whom were no longer in business. It was thus inevitable that the rather uninformed design decisions which had to be made resulted in a number of titanium tube failures from vibration and fretting fatigue. The problem is now resolved in the ninth edition of the Heat Exchange Institute *Standards for Surface Condensers* [HEI 1995].

Another area not fully addressed in early retubing was the tubesheet material. Almost all condensers receiving seawater or brackish water are provided with tubesheets made from brass- or bronze-copper alloy. With titanium tubing, these cooling waters act as an electrolyte and create a galvanic action between the less noble copper alloy tubesheet and the more noble titanium, leading to a preferential corrosion of the copper alloy. If this condition is ignored, it could lead to eventual perforation of tubes as well as leaking tube joints, threatening the inherent integrity of the titanium tubes. In order to reduce the possibility of leaks from this source, the tubesheets and, in some cases, even the waterboxes were coated with an epoxy film. Unfortunately, this coating has to be replaced periodically. Current practice where titanium tubes are to be used is to design condensers with tubesheets also made from titanium. If solid titanium tubesheets are specified, then tube joints can be roller-expanded; and if the ultimate in joint tightness is desired, the tube joints can always be welded after rolling. Further, the "all-titanium" condenser is carefully designed to avoid the effects of high steam velocity on tube vibration, the most common cause of failure with brass tubes.

The protective film on titanium tubes is a tenacious oxide which, if it becomes damaged, re-forms immediately in the presence of moisture. It is also impervious to the appearance of smeared iron on the tube surface. Smeared iron becomes harmful only at the critical temperature of 170°F, much higher than the temperatures at which condensers normally operate and therefore unlikely to occur. Thus, titanium tubes can be cleaned using carbon steel cleaners and brought back to a "new" condition without concern.

Titanium has many advantages over other metals. For instance, if a titanium tube becomes partially plugged with stones or shellfish, the tube will not suffer impingement attack or deposit corrosion. Titanium tubes cannot environmentally contaminate cooling-water effluent, or become corroded from ammoniated condensate. In addition, the condensate cannot become contaminated from metal ions. In fact, titanium tube samples removed from condensers after 20 years of service have been found to be like new. However, even titanium tubes can become fouled, so that condenser performance falls below its design heat transfer capacity—although this can be restored by periodic cleaning.

11.3 MECHANICAL CLEANING SYSTEMS AND TOOLS

The earliest mechanical tube cleaners were designed for cleaning boiler tubes. It was not until 1923 that two brothers, Cecil M. Griffin and Vivian Griffin, invented the first cleaner for condensers and shell-and-tube heat exchangers. Cecil Griffin's initial patent was granted in 1931 [Griffin 1931], and subsequent patents which improved on the original design were granted in 1939 [Griffin 1939]. Still further improvements occurred in 1947 [Griffin 1947] and 1956 [Griffin 1956]. Figure 11.4 (a) shows a current version of this cleaner, which consist of several U-shaped tempered steel strips arranged to form pairs of spring-loaded blades. These strips are mounted on a spindle and placed at 90° rotation to one another. Mounted at one end of the spindle is a serrated rubber or plastic disk (developed by George Saxon [1981]) which allows a jet of water, delivered by a pump operating at 300 psig, to propel the cleaners through a tube with greater hydraulic efficiency. Figure 11.5 (a) shows a typical pump, mounted on a portable skid, complete with the handheld triggered device (Figure 11.5 (b), also known as a gun), which permits the operator to direct the water to the tube being cleaned.

Since the earlier cleaner designs behaved as stiff springs, loading the cleaners into the tubes was sometimes rather tedious. To speed up this operation, while also providing the blades with more circumferential coverage of the tube surface, the hexagonal cleaner shown in Figure 11.4 (b) was developed by Saxon and Krysicki [1998]. This design not only reduced the cleaning time for 1000 tubes

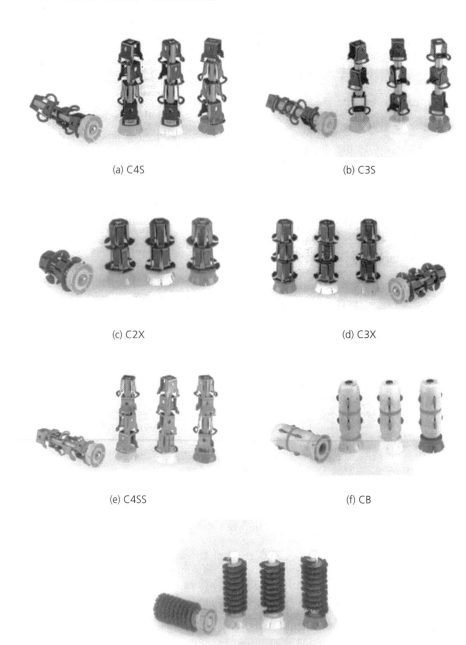

(a) C4S

(b) C3S

(c) C2X

(d) C3X

(e) C4SS

(f) CB

(g) H-Brush

FIGURE 11.4. Mechanical Cleaners—(a) C4S, (b) C3S, (c) C2X, (d) C3X, (e) C4SS, (f) Calcite cleaner, (g) brush (Source: Conco Systems, Inc.).

(a) Portable Booster Pump

(b) Water Gun

Figure 11.5. Water pump and gun for mechanical cleaners (Source: Conco Systems, Inc.).

but also was found to be more efficient in removing tenacious deposits such as those consisting of various forms of manganese.

A later development by Gregory Saxon [1992] involved a tool for removing hard calcite deposits, which were found to be difficult to remove even by acid cleaning. This tool is also shown in Figure 11.4 (c), and consists of a Teflon body on which are mounted a number of rotary cutters, placed at different angles and provided with a plastic disk similar to those used to propel other cleaners through tubes. Used on the condenser discussed later in this chapter in the first of the case studies, cleaners of this type removed 80 tons of calcite material. The tool has now become standard whenever hard and brittle deposits are encountered.

Mechanical cleaners offer the most effective off-line tube cleaning method. Strong enough to remove hard deposits, they can cut the peaks surrounding pits and flush out the residue at the same time, thus retarding underdeposit corrosion. These cleaning tools are not offered in "one size fits all" form. Each is custom-made to fit snugly inside a tube having a stated diameter and gage or wall thickness. This allows the blade contact pressure to be controlled within tolerances and also ensures that as much of the tube circumference as possible is covered as the cleaner passes down a tube of the corresponding size.

Reference was made in Section 1.4 to the use of a special rig for the heat transfer testing of tubes. In the development of these various cleaners, extensive use was made of this rig to determine the effectiveness of different cleaner designs. Not only could the incremental material removed be established for each successive pass of the tool, but the cleaner's effect on heat transfer could also be quantified for a given kind of deposit.

In some plants, it has been a practice to use compressed air with water to propel the cleaners. However, the rapid expansion of the air when the cleaner exits from the tube causes the cleaner to become a dangerous projectile; whereas the use of water alone never allows the cleaner exit velocity to rise above a safe level. Similarly, water supplied at a pressure of 300 psig is much safer to personnel than at up to 10,000 psig, which is sometimes attained when using high-pressure hydrolasing techniques. Another advantage of mechanically cleaning condenser tubes using water as the cleaner propellant is that the material removed can be collected in a plastic container for later drying and weighing to establish the deposit density (in grams per square foot). In many cases, X-ray fluorescent analysis of the deposit cake is performed. Experience has also shown that properly designed cleaners will not become stuck inside tubes, unless the tube is already damaged or severely obstructed.

The concern is occasionally expressed that mechanical cleaners can possibly cause damage to tube surfaces. Damage is extremely rare with cleaners which have been properly designed and carefully manufactured. Indeed, Hovland et al. [1988] conducted controlled experiments by passing such cleaners repeatedly through 30-ft-long 90-10 Cu-Ni tubes, adhering to the cleaning procedures normally used on-site, and then measuring any change in wall thickness. After 100 passes of these cleaners, the wall thickness was reduced by only 0.0005 to 0.0009 in. If a 50% reduction in wall thickness is a critical parameter, extrapolating the reduction in wall thickness experienced during this series of tests would be equivalent to 2800 passes of a cleaner per tube. Considering that condensers are seldom cleaned more than twice per year, the wall life reduction due to cleaning can be spread over more than 1000 years!

Hovland et al. [1988] also showed that discarding a cleaner after it had been used 10 times reduced the wall-thickness reduction. Further, the largest wall-thickness reductions were found to occur on new tubes during the first two passes. This can be attributed to these initial passes functioning as smoothing or cleanout passes, in which tube irregularities and foreign material are removed together with an "exfoliation" of the base metal. Later passes show a much lower, and still minuscule, rate of material removal.

Establishing the optimal frequency of condenser cleaning was discussed in Chapter 7. However, with nuclear plants, there is less flexibility, and it is considered good practice to clean at every refueling outage. Even where a nuclear unit is equipped with an on-line cleaning system, an annual or biannual off-line mechanical cleaning is good practice and (1) assures that the condenser effectiveness will be maintained, (2) reduces the risk of pitting from the stagnant water in those tubes which may become blocked by stuck sponge balls, and (3) ensures that every tube is cleared and cleaned at least once or twice a year.

11.3.1 A Family of Metal Cleaners

As already mentioned, Cecil M. and Vivian Griffin, engineers with Duquesne Light Company in Pittsburgh and also brothers, invented the first mechanical tube cleaning system for condensers [Griffin 1931] during the early 1920s. In 1923, the Griffin brothers, both graduates of what is now Carnegie-Mellon University, formed their own company to manufacture these tools. As has been noted, Cecil Griffin was awarded several later patents [Griffin 1939, 1947, 1956]. Their company was taken over in 1971 by George E. Saxon, Sr.,* and since then has prospered and flourished. The development of this technology has continued with the introduction of several new devices, each specially designed to tackle one of the harder deposits to remove.

Spring-loaded carbon steel cleaners have long been recognized as the most effective tube cleaner available to the industry. They come in various forms, designed for multideposit removal, for the removal of hard deposits, or for removal of thin and tenacious deposits. Since the blades of these cleaners are spring-loaded with a determined force at the surface of contact, they can be engineered to perform the intended function while also improving the smoothness of the tube surface during cleaning as they restore the tube to its "as new" condition.

Mechanical cleaners travel through the tubes at a velocity of 10 to 20 ft/s and are propelled by water delivered at 300 psig. Some of the members of a well-known family of tube cleaners have the following features:

- *C4S.* The C4S cleaner is a general-purpose type and may be used to remove all types of obstructions, deposit corrosion, and pitting.
- *C3S.* The C3S cleaner is designed for heavy duty and is very effective in removing all kinds of tenacious deposits. Its reinforced construction also allows it to remove hard deposits, corrosion deposits, and obstructions.
- *C2X, C3X.* These types of cleaner consist of two or three hexagonal-bladed cleaner elements, having six arcs of contact per blade. They are effective on all types of deposits but are especially suitable for removing the thin tenacious deposits of iron, manganese, or silica found on either stainless steel, titanium, or copper-based tube material.
- *C4SS.* While the C4SS stainless steel cleaners can be used with all types of tube materials, they were originally developed for applications using AL-6XN stainless steel condenser tubes but have also been used for cleaning tubes in highly corrosive environments.
- *CB.* This tube cleaner was specifically developed to remove hard calcium carbonate deposits and is designed to break the eggshell-like crystalline form characteristic of these tube scaling deposits. This cleaner has been

---

*It is now known as Conco Systems, Inc., and is located in Verona, Pennsylvania.

found exceptionally helpful in avoiding the need for alternative and environmentally harmful chemical cleaning methods, which were all that were previously available for removing these hard deposits.

11.3.2 Brushes

Brushes are principally intended to remove light organic deposits such as silt or mud. They are also useful for cleaning tubes with enhanced surfaces (e.g., spirally indented or finned), or those with thin-wall metal inserts or tube coatings. The brush length can be increased for more effective removal of lighter deposits, as illustrated in Figure 11.4 (d).

11.3.3 Cleaning Productivity in the Field

Using the methods outlined earlier in Section 11.3, it is possible for between 5000 and 7000 tubes to be cleaned during a 12-hour shift, utilizing a crew of four operators with two water guns supplied from one pump. Clearly, an increase in crew size, in the number of water guns available, or in the number of mechanical cleaners supplied for the project can increase the number of tubes which can be cleaned during a shift, provided there is adequate space in the waterbox(es) for the crew to work effectively.

11.3.4 Some Limitations of Mechanical Cleaning Methods

Each type of off-line cleaning device has its own limitations. For instance, brushes may be effective only with the softest fouling deposits, whereas metal cleaners and high-pressure water are more effective against tenacious foulants. However, all methods may need assistance where the deposits have been allowed to build up and even become hard. In such cases, it may still be necessary to acid-clean, followed by cleaning with mechanical cleaners or high-pressure water to remove any remaining debris.

11.4 FOULING DEPOSIT CHARACTERISTICS

Section 10.1 listed some of the most frequent causes of fouling:

- Sedimentary fouling or silt formation
- Deposits of organic or inorganic salts
- Microbiological fouling
- Macrobiological fouling

However, in the course of cleaning condensers not only in the United States but also in most industrialized nations, at least 1000 different types of fouling deposit have been encountered. Each power plant has its own fouling idiosyncrasies, and even in one plant, it is not uncommon to find that the fouling characteristics are different for each unit, even when the condenser tubes are of the same material and the equipment was built to the same set of drawings. It is also not uncommon to find that fouling of the bundles made from copper alloy tubes is of a different nature from fouling in the stainless steel bundles which are frequently associated with the air-removal section in the same condenser.

After retubing a condenser, it is also quite normal for the fouling characteristics to be significantly altered. For instance, the types of deposit encountered with stainless steel tubes (especially the manganese problem outlined in Section 10.3.1) are not the same as those found on the inside surfaces of admiralty tubes in the same location. Should a condenser be retubed with titanium tubes, then the fouling and corrosion problems will again be different from those previously experienced.

11.4.1 Particulate Fouling

Particulate fouling occurs when small particles of matter attach themselves to the tube wall. Generally, there must already be a substrate of deposit material to which the particulates become attached, the common substrate being bacterial slime. Once attached, the deposit layer accumulates rapidly with a corresponding increase in thermal heat transfer resistance. The standard tube velocities of around 7 ft/s are often high enough to remove new material at the same rate at which it becomes deposited, creating an equilibrium condition during which no further deterioration in performance occurs. This fouling process has been termed "asymptotic fouling." Off-line cleaning methods are commonly used to remove particulate fouling.

11.4.2 Crystalline Fouling

Examples of crystalline fouling frequently encountered, in descending order of thermal resistance, are microbiological (see Section 11.4.3), manganese, iron, silica, calcium carbonate and calcium phosphate. Crystallization occurs when the solubility of the salts in question is exceeded; the surplus salts come out of solution and become deposited on the surfaces of the tubes. Some salts exhibit an *inverse* solubility in which the solubility *decreases* with an increase in temperature. This is especially true of calcium carbonate, and it is common to find minimal fouling on the circulating water piping and waterboxes but an increase in fouling toward the outlets of the tubes. The removal of calcium carbonate is a difficult task and has required the development of the special tools described by Saxon [1992].

11.4.3 Microbiological Fouling

Bacteria exist in all sources of cooling water available to power plants. Many of these bacteria exude a gelatinous substance which assists the bacteria in attaching themselves to the tube walls; because these slimes contain some 90% water, they have a high resistance to heat transfer. Thus, slime-forming bacteria can have a serious negative impact on condenser performance, and, unlike the slow impact of crystalline fouling on tube heat transfer, microbiological fouling is immediately detected through a rapid falling off in condenser and unit performance.

11.4.4 Macrobiological Fouling

Macrofouling can occur with either fresh or sea water, so that any power station is liable to find macrofouling in condenser cooling-water and/or nuclear plant service-water systems. Fresh-water fouling by Asiatic clams, zebra mussels, and other shellfish will seriously reduce the water flow rate through a condenser, but good maintenance practices can help to control it. Of course, where organisms cause tube blockage, the reduced flow can result in an increase in sedimentation and, if it is allowed to accumulate, an increase in corrosion of both condenser and heat exchanger tubes.

When zebra mussels enter a plant through the intake screens, they are usually in the larval stage and settle in areas of low velocity (e.g., below 6.5 ft/s), where they can grow to 1.25 inches when mature. Zebra mussels are able to attach themselves to any surface, including other mussels, and can sometimes form a mat which can be 4 inches or more thick. Adult mussels attach themselves by means of byssal threads and can grow as many as 12 new threads a day. It is this ability to attach and form colonies or masses that causes the zebra mussel to be, potentially, a major threat to plant operation and reliability. Meanwhile, since Asiatic clams find it difficult to attach themselves to the tube walls, they also tend to be found in areas of low velocity, where they can grow and reproduce.

Aggressive cleaning of both the tubes and tubesheets is necessary to control macrofouling, as was the case in the plant described in Section 6.2.

11.4.5 Corrosion Product Fouling

Copper alloys develop a relatively thick protective cuprous oxide film, although the thermal conductivity of the tubes will decline until the protective layer has been fully developed. Where the water is "aggressive," or in cases of localized corrosion, copper oxide deposits can grow to a considerable thickness. Eventually, this type of copper corrosion will affect the heat transfer capability of the tubes significantly and must then be removed.

11.5 DEPOSIT SAMPLING

Analysis of the deposits removed from the main condenser is an important part of selecting the proper cleaning process. When access to the condenser is possible during reduced-load operation or a unit outage, deposit samples can be obtained from the whole length of a tube as well as from tubes in several areas of the tubesheet. This is important in that fouling is seldom distributed uniformly throughout the condenser, because of flow and temperature variations. Thus, it is important to have a sampling plan in place for use when access to the condenser becomes possible. It has also been found that, when using water at 300 psig to propel the cleaners, deposit samples can be collected from all tube locations within any waterbox, without damage to equipment or danger to personnel.

The dry weight per unit area of the deposits collected from a tube and their composition, as determined by X-ray fluorescent analysis, can be correlated with changes in condenser performance based on historical operating data. The distribution of deposit intensity within tubes and across the tubesheet or waterbox can also assist in developing an appropriate cleaning strategy, including cleaning frequency or water treatment procedure. If tube samples in the fouled condition can be made available for heat transfer testing, the expected improvement in heat transfer from a selected cleaning procedure, or a set of alternative procedures, can also help in providing a quantifiable prediction of return on investment.

11.6 DEVELOPING AN APPROPRIATE CLEANING PROCEDURE

Regardless of the material, the most effective way to prevent tubes from deteriorating before expected life runs out is to keep them clean. Removing tube deposits, sedimentation, biofouling, and obstructions and returning the tube surface almost to bare metal is important for renewing the tube metal life cycle, recognizing that the protective oxide coatings will quickly rebuild themselves and passivate the cleaned tube. Whatever the type(s) of deposit present, the cleaning procedure should also be designed to remove them as completely as possible while also minimizing the time the unit is out of service. Simply going through the cleaning process without evaluating the available information can become a futile exercise. Thus, adopting the most effective cleaning process will tend to yield the best results; some of the issues to consider are described in the following.

11.6.1 Removal of Obstructions

Obstructions within tubes, as well as many forms of macrofouling, can so obstruct the flow of the circulating water that they make many tube cleaning methods ineffective; clearly, when obstructions are so severe, those methods should be avoided. Attention has already been drawn to obstructions by shellfish, including Asiatic clams and zebra mussels (i.e., macrofouling); a tube cleaner must have the body and strength to remove such obstructions. The cleaning method must also be able to remove the byssal material with which shellfish attach themselves to the tube walls. Certain other types of debris can also become obstructions, among them cooling tower fill, construction waste, sponge rubber balls, rocks, sticks, twigs, seaweed, and fresh-water pollutants, any or all of which can become lodged in the tubes and will then have to be removed.

11.6.2   Removal of Corrosion Products

As already mentioned, copper deposits grow continuously, and the thick oxide coating or corrosion product can grow to the point where it will seriously impede heat transfer. Not only will the performance of the condenser be degraded, but such deposits will also increase the potential for tube failure. When a thick outer layer of porous cupric oxide is allowed to develop, it disrupts the protective cuprous oxide film, allowing the base metal to be attacked and underdeposit pitting to develop. Such destructive cupric oxide accumulations must be removed regularly, together with any other deposits.

11.6.3   Surface Roughness

Rough tube surfaces are associated with increased friction coefficients and reduced cooling-water flow rates, the latter allowing deposits to accumulate faster. It has also been found that rough tube surfaces tend to pit more easily than smooth surfaces. Smooth tube surfaces can improve condenser and heat exchanger performance in at least five ways:

- Reduced heat lost to the environment and improved heat transfer capacity. The smoother tube wall reduces the operating temperatures while also increasing the circulating-water flow rate. The increased flow volume and velocity will also result in a lower water temperature rise across the condenser.
- Reduced pumping power. The pressure drop across the condenser is also lower, thus lessening any tendency for the discharge vacuum to become broken and interfering with the siphon effect. This also helps to ensure that water is always flowing through the top rows of tubes.
- Increased time can be allowed between cleanings because of the reduced rate of redeposition of fouling material on the tube surfaces.
- Reduced pitting as a result of turbulence and gas bubble implosion. Mechanical cleaning will cut off the peaks of the pits while, at the same time, the water will flush out the corrosion products. This retards future pitting and increases tube life. Tenacious deposits and tube obstructions are also much easier to remove if the surfaces of the tubes are smooth.
- Smooth tube surfaces also enhance the results obtained from eddy-current testing, extend probe life, and increase signal reliability, all further improving plant availability.

11.7 CASE STUDIES

This section contains some case studies of plants which have used mechanical tube cleaning not just to improve unit heat rate, but also to regain generation capacity lost to fouling.

11.7.1 Case Study 1: Clinton Power Station

Background

Plant performance at the Clinton Power Station in Illinois had degraded over four operation cycles. The performance evaluation program at the station had monitored the thermal performance of the unit and recommended ways to improve it, with the goal of cost-effective optimization of electrical power production. The program identified condenser-related thermal deficiencies. Over time, the magnitude of the deficiencies increased, resulting in seasonal output reductions of up to 15 MW due to elevated back pressure [Saxon and Putman 1996]. As the back-pressure data was monitored and trended, plant personnel became concerned when it became 0.8 in-Hg higher than expected. Two years later the deviation had peaked at 1.3 in-Hg. Hard scale deposits were known to be present as a result of analysis and inspection.

Condenser Physical Data

The generating unit at Clinton Power Station is a 985-MW boiling-water reactor (BWR). The condenser is of the single-shell, single-pass type with a divided waterbox arrangement. There are 53,160 tubes, with dimensions 7/8 in outside diameter, 22 BWG, and 70 ft long, made from 304 stainless steel. The unit has a closed cooling system drawing fresh water from an artificial lake.

Performance Analysis

The performance analysis showed a serious deviation in back pressure. There were also dramatic swings in the pH of the lake water, making it extremely difficult to manage the required scale inhibitor feed rate. Although previous mechanical cleanings had effectively removed soft deposits, hard deposits were still present.

The numbers entered in Table 11.3 are the averages of losses due to degraded condenser performance under the highest and lowest lake temperatures for each month. The generation revenue loss calculated at $17/MWh is in excess of $2.8 million. These three sets of data are for the expected generation output losses at design and at off-design conditions [Stiemsma 1994].

During an examination of the tubes, scale deposits were measured at 20 mils, and calculations established a heat transfer resistance that would result in an

Table 11.3. Clinton Power Station Megawatt Loss Due to Operating above Design Back Pressure

Month	Design	MW Loss Due to Operation At	
		0.6 in-Hg Above Design	1.3 in-Hg Above Design
January	0.0	0.0	2.41
February	0.0	0.0	2.66
March	0.0	0.0	3.35
April	0.0	1.12	6.05
May	0.0	1.56	7.60
June	1.74	6.66	14.47
July	3.20	9.53	19.96
August	3.49	10.04	20.56
September	1.06	5.39	14.45
October	0.0	1.39	7.46
November	0.0	0.14	4.12
December	0.0	0.0	2.26

elevation of condenser back pressure which, it was believed, could account for 40% of the observed difference. Deposit analysis proved the deposit to be calcium carbonate scale. A tube sample was pulled from the unit and cut into eight sections, the heat transfer test results for Sections 2, 4, 6 and 8 being shown in Table 11.4.

Development of a New Cleaner for Ceramic Deposits

Although attempts were made to remove the deposits using conventional scraping methods, it was decided that a new type of mechanical scale cutter had to be developed, based on a completely different technique [Saxon 1992], to remove this ceramiclike deposit. The new cutters adopted had features similar to common glass cutters and were designed to fracture the scale. Used in combination

Table 11.4. Test U Coefficient for Clinton Power Station Tube Samples, Btu/ft^2 · h · °F)

Condition	Section 2	Section 4	Section 6	Section 8
As found	383.03	376.86	367.32	337.48
Mechanically cleaned	521.50	554.85	597.39	583.89
Acid-washed	515.28	554.93	593.07	582.15

with conventional scrapers, the new cutters proved effective in experiments carried out at the manufacturer's laboratory and were applied during the upcoming outage.

Results

Following the development of the new cleaning technology and its application in the field, 80 tons of scale were removed from the condenser tubes. A subsequent boroscopic examination revealed that this amount of calcite material constituted approximately 75% of the scale which had been present in the tubes prior to cleaning. Condenser performance was returned to the best level ever; had the condenser not been cleaned, a generation revenue loss of at least $2.6 million would have occurred during the next operation cycle [Stiemsma 1994].

11.7.2   Case Study II: Peach Bottom Atomic Power Station, Pennsylvania

The Peach Bottom plant had typically monitored the cleanliness factor as the principal performance criterion for determining the presence of deposits and obstructions in the condenser tubes. Eventually, thermal performance engineers began to observe a degradation in the cleanliness factor extending over several months of operation [Bell 1994]. The performance degradation in unit 2 was thought to be due to debris blocking the tubes. However, the original observations indicated that degradation was occurring even during the winter months, when the cooling-water temperatures are normally low. As a result, the performance loss became of greater concern, since some fouling of the condenser ought to be possible during the winter without an accompanying megawatt output loss. One factor considered was that the hyperchlorite system the plant used to control biological growth also promoted manganese plateout.

Condenser Physical Data

The Peach Bottom unit is a 1152-MW boiling-water reactor, with a single-pass condenser of the once-through type with six waterboxes and drawing cooling water from the Susquehanna River. The condenser contains 55,080 titanium tubes, whose dimensions are 1 in outside diameter, 22 BWG, and 50 ft long.

Performance Analysis

As already mentioned, even with colder cooling-water temperatures, the reduction in cleanliness factor was enough to produce an output loss [Bell 1994]. However, it was possible to clean the waterboxes during a load drop in February and this resulted in the recovery of 6 MW, while the cleanliness factor also improved,

Table 11.5. Dry Weights and Deposit Information from PAS-2 Whole-Tube Deposit Samples from the Peach Bottom Plant

Dry Weight, g	Density, g
217.30	17.5945
129.89	10.5170
126.12	10.2118
111.10	8.9956
98.62	7.9851
95.56	7.7374
93.54	7.5738
36.37	2.9448
25.26	2.0452
4.95	0.4007
1.79	0.1449

but not significantly. A plan was then devised for deposit sampling. Mechanical tube cleaners were propelled through the tubes for whole-tube deposit recovery and the elements present were identified and quantified. Deposit weight densities collected from various locations in the condenser ranged form 0.4 to 17.5 g/ft^2 over 11 samples taken, as shown in Table 11.5. The deposit sampling confirmed the uneven distribution of fouling throughout the condenser, while the elemental findings confirmed suspicion of manganese plating, manganese also being predominant in the Susquehanna River. The elemental analysis findings are given in Table 11.6.

New Cleaner Development

Standard C4S or spring-loaded four-bladed cleaners were used in the deposit sampling. Internal tests were performed on blades having various tensions, and the surface contact was increased in the search for the optimal method for removing the Peach Bottom deposits. The presence of the manganese plating, and the knowledge that even more severe problems with manganese fouling existed at other plants [Anderson 1990], stimulated the search for a new technology for removing manganese deposits. The improved cleaner design which resulted [Saxon 1998] still further increased surface contact area and provided more effective removal of thin tenacious deposits such as manganese.

Results

55,000 condenser tubes were cleaned in 90 hours during the 10th refueling outage of Peach Bottom's unit 2. As a result of those four days of work, the plant

Table 11.6.   Results of Elemental Analysis—Peach Bottom
Plant

Element	%
Manganese	10–20
Aluminum	0.1–1.0
Potassium	0.1–1.0
Iron	5–10
Phosphorus	0.1–1.0
Titanium	0.1–1.0
Silicon	5–10
Sulfur	0.1–1.0
Nickel	0.1–1.0
Calcium	1–5
Chlorine	0.1–1.0
Elements <0.01%	Not listed
Loss on ignition	Not listed

operating company was expected to recoup lost energy amounting to 25 MW generating capacity, equivalent to more than $4.3 million per year [Bell 1994]. The tube cleaners removed 7000 pounds of deposits, and the condenser achieved peak performance when it was returned to operation.

11.7.3   Case Study 3: Power Plant in Trinidad and Tobago

A 67-MW combined-cycle plant in Trinidad and Tobago was also experiencing limitations on gross megawatt output due to poor condenser performance. It was able to generate only a maximum of 30 MW, less than half its original potential. The condenser had approximately ten thousand 3/4-in OD stainless steel tubes which were observed to be fouled. An inspection and a cleaning program were prepared for implementation during an upcoming unit outage. As a result, an additional 34 MW of capacity was recovered, at a back pressure of just over 3 in-Hg, providing an improvement of 0.67 in-Hg. In all, some 6000 pounds of deposits were removed from this condenser.

11.7.4   Case Study 4: San Diego Gas and Electric

A 300-MW unit operated by San Diego Gas and Electric had condenser fouling problems which were affecting heat rate and, at times, generation capacity as well. In 1965, a section of the condenser was retubed, but after the unit was brought back on-line, it was found that the back pressure had fallen only from 2.3 in-Hg before the partial retube to 2 in-Hg afterward. Further, the back pressure still continued to rise, having its usual negative effect on unit heat rate.

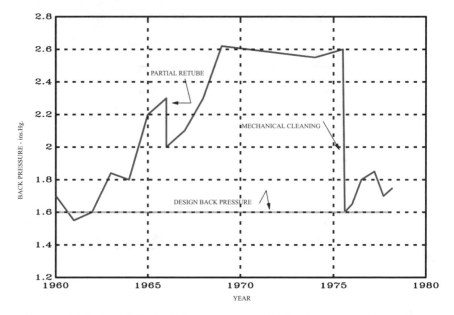

FIGURE 11.6. Case study 4: improvement in back-pressure deviation resulting from partial retubing vs. mechanical cleaning.

The 1-in OD, 18 BWG tubes in this condenser were made from aluminum brass, cooled using seawater, and had been treated with ferrous sulfate in order to passivate the tube surfaces. Some buildup of ferrous sulfate deposits had occurred. In 1975, the plant decided it had to master these fouling problems by thoroughly cleaning the whole condenser using mechanical cleaners. A total of about 7000 lb of deposits, mostly ferrous sulfate, were removed during this process. Hovland (1978) shows in Figure 11.6 the remarkable improvement which resulted, the back pressure now falling from its persistent value of 2.6 in-Hg to its design value of 1.6 in-Hg. This was equivalent to at least a 1% improvement in heat rate at full load, or $350,000 a year.

11.7.5 Conclusions

Effective diagnostic procedures can lead to an accurate assessment of a power plant's problems with condenser fouling and corrosion and to innovative cleaning solutions well suited to the situation. Case studies 1 and 2 both illustrate the development of new cleaning technologies to deal with current fouling problems (the new technologies were subsequently applied to other sites that had similar problems, with similar positive results). Successful diagnosis and resolution of condenser fouling problems is largely based on the ability to identify the particular type of fouling involved and the most appropriate cleaning technology. Since the nature of the deposits varies from site to site, the key to effective and successful cleaning is to marry the characteristics of the deposit to the cleaning technique appropriate to the site. In the new and highly competitive utility industry environment, any loss of generation capacity, or any performance problems allowed to persist into future operating cycles, can be very costly. The improvements in output capacity which can result from coming to grips with the situation can be dramatic, as exemplified in all the case studies.

11.8 WATERBOX SAFETY PROCEDURES

In order to clean condenser tubes, service personnel need to enter the waterboxes to gain access, and they must also be able to work efficiently and safely in these confined spaces. However, by their nature, such spaces are considered to be hazardous working situations, and proper safety procedures have to be faithfully followed. Further, since the details of waterbox construction differ, the procedures must be defined for each waterbox before workers enter it; the procedures for the inlet waterbox may be different from those for the outlet waterbox on the same condenser.

The Occupational Safety and Health Administration (OSHA) standard for permit-required confined spaces (29 CFR 1910.146) [OSHA 1999] applies and became effective on April 15, 1993. OSHA defines a confined space as:

- Is large enough and so configured that an employee can bodily enter and perform assigned work; and
- Has limited or restricted means for entry or exit (for example, tanks, vessels, silos, storage bins, hoppers, vaults, and pits are spaces that may have limited means of entry); and
- The space is not designed for continuous worker occupancy.

The openings are usually small and are awkward for people to pass through easily. Their small size also makes it difficult to move needed equipment into the space, including respirators or lifesaving equipment. However, although a confined space may have a limited or restricted means for entry and exit, it still has to be large enough for personnel to work within it. Ventilation is usually poor, the atmosphere often being oxygen-deficient. The range of oxygen concentration considered safe for entry is when the atmosphere contains more than 19.5% and less than 23.5% oxygen.

Clearly, a confined space is not suitable for continuous employee occupancy and, since confined spaces expose personnel to hazards, special precautions have to be taken to prevent serious injury or, indeed, loss of life. The rules for a permit require that a confined-space safety program, documented in writing, be in place in every plant where equipment with such spaces (condensers normally fall into this category) may require entry. The OSHA standard also insists that every authorized entrant into a confined space be supervised by an entry supervisor and that an attendant be stationed outside the confined space for the whole of the time it is occupied. The attendant has the duty and responsibility to monitor the status of the entrants and to alert them should there be any need to evacuate the space. Rescue and emergency procedures must also be defined and personnel available on call.

The cleaning of condensers or heat exchangers is normally a team effort with many people working side by side inside the waterbox. Thus, to reduce the exposure to risk of all personnel working in these confined spaces, *all* must receive the most thorough training so that they are made aware of the range of possible situations which can occur and can support each other as a team, especially during stressful periods. Several training programs are available, a typical text being Keller [1992]. The training topics should include the following:

- General safety
- Emergency response and evacuation procedures
- Training for working within confined spaces
- Protecting oneself from accidental slips and falls
- CPR/first aid training
- Training in scaffold construction and inspection
- Lockout and tagout procedures
- Use of air monitoring equipment and ventilation provision
- Low-voltage lighting and ground fault interceptor equipment
- Personal equipment which must be worn or kept handy

To ensure safety and prevent injuries while working, it is necessary to be properly dressed and equipped. Thus, hard hats, safety glasses, eye and ear protection, safety-toed footwear, coveralls, and harnesses are obligatory. Neglecting to wear all these articles can also lead to severe legal penalties. All power-plant workers must be responsible for their own actions, and, besides working smart, all must make a practice of following more than the minimum requirements if the safety of the whole team is to be ensured.

Prior to the cleaning activity, the work area should first be secured and taped off with caution tape. Then, when entering a waterbox, care must be taken to ensure that it has been completely drained. Becoming unexpectedly submersed in water while within the confined space could cause panic and possibly lead to an accident. There is also the possibility that the atmosphere within a waterbox can become dangerous: perhaps the oxygen concentration has dropped to unsafe levels because of inadequate ventilation, or toxic or explosive gases, fumes, or vapors may have accumulated within it. The OSHA standard lays down requirements for testing the atmosphere in the confined space before entry is permitted and to determine whether some form of forced ventilation will be required.

Waterboxes tend to be rather wet environments, and, even with the utmost care, the presence of lighting or other electrical equipment can lead to accidental exposure to electrical malfunctions. Even voltages as low as 110 V can lead to electrocution if proper precautions are not taken. There are also several other high-energy sources within a power plant, including water under pressure and pneumatic and hydraulic systems. In all cases, the lockout and tagout procedures must be the first line of defense, including tagout of valves: should a cooling-water inlet valve be opened accidentally, a waterbox can quickly become flooded and engulf anyone working inside. Workers should also be knowledgeable of the *air horn*, which is provided as a means of reducing the ambient temperature in these confined spaces. If not properly used, this 10-pound metal object attached to an inexhaustible air supply can cause injury.

There are also potential biological, mechanical, and chemical hazards. Any water not drained out, and to which the crew can therefore become exposed, may contain bacteria which can infect cuts or bruises if these are not properly treated. The barnacle buildup on the walls of waterboxes can have very sharp edges, and people can be cut accidentally by cathodic protection probes while concentrating on the task at hand. Other types of cooling-water bacteria can cause disease (e.g., Legionnaires' disease), a possibility operators should be aware of. Personnel should also be aware of any unusual chemicals, unpleasant or dangerously hot areas (e.g., boiler casings or high-pressure steam lines), or high-decibel noise sources in the vicinity of the work area.

A number of safety issues surround *scaffolding*. Waterboxes can be so large that scaffolding is necessary for reaching and cleaning some of the tubes. Scaffolding is any rig or platform one can stand on above floor level to extend one's reach. In a waterbox, it is likely to consist of planks temporarily placed on brackets or fixtures permanently attached to the walls; the planks must be secured to the fixtures with number 9 wire or rope. In the absence of any better means, planks may be prevented from tipping by placing rods of an appropriate size in the tubesheet. If mounting the scaffold is a problem, an access ladder may have to be installed. After the scaffolding has been properly erected, hoses and electrical extension cords should be run overhead so that people will not become entangled in them.

Care must be taken that the design and construction of the scaffolding are safe. The number of planks should depend *only* on how many are required to ensure safe working conditions; there should be no attempt to economize. The planking should also be fitted with toeboards, to reduce the chance for tools and other objects to fall off and injure people working below. Improperly secured scaffolding can lead to physical injury through slips or falls. The scaffold brackets permanently installed within the waterbox must be regularly inspected for corrosion or weakening. The design should ensure, as far as possible, that people will be unlikely to slip on the possibly wet surfaces which are commonly found in this working environment. Above all, if a scaffold *feels* unsafe, it must be repaired or reinforced before work can begin or continue.

Personnel working within the waterbox have their own responsibility for making the scaffolding safe. When mounting the scaffold, they should correctly use the access ladder, if provided; because of its sharp edges, cathodic protection equipment should never be used as hand- or footholds. No unnecessary item should be taken onto the scaffold. Finally, it is unacceptable and bad practice to attempt to use unsecured objects such as an upturned bucket to gain better tube access.

For summoning help in an emergency, all members of the team must know the location of the nearest plant telephone, as well as the set of important numbers they may need, this information being included on the confined-space entry permit. In conclusion, all workers must work together as a team, each following only the safest practices at all times, to ensure their own and all their fellow workers' safety. Thorough training and good practice are key, coupled with careful planning of the cleaning process and rehearsed evacuation procedures in response to all anticipated emergencies.

Chapter 12

AIR AND WATER INLEAKAGE DETECTION AND EDDY-CURRENT TESTING

12.1 INTRODUCTION

Many different methods are available for detecting condenser water and air inleakage, some based on the chemical analysis of the condensate and others monitoring the quantity of air entrained in the vapor/gas mixture ejected by the air-removal system. However, once air or water inleakage is detected, it is necessary to locate all sources (there may be more than one) and then rectify them so that the unit's performance can be restored or the quality of the feedwater safeguarded.

The nature and severity of a leak largely determine the best method for locating its source. Tracer gases such as helium and SF_6 have become widely accepted over the past 10 years as the standard operating procedure. In addition, the isolation of leaks in condenser tube bundles by on-line injection of SF_6 into the circulating-water inlet has become state-of-the-art. Tracer gas technology has eliminated much of the guesswork normally experienced when using less sophisticated methods and has resulted in shorter downtimes required to perform leak detection. It has greatly helped to reduce the increase in heat rate associated with air inleakage. This chapter will describe the practical application and use of tracer gases in leak detection and will compare it with earlier and less sophisticated techniques. Selection of the most appropriate procedure and tracer gas in a given application is also discussed. The final sections of the chapter offer detailed discussions of eddy-current testing and the plugging of tube leaks.

12.2 INTUITIVE METHODS OF LEAK DETECTION

Prior to 1978, among the techniques used for condenser tube water inleakage inspections were shaving cream, sheets of plastic, smoke generators, sight, and hearing. None of these techniques proved to be very reliable. Cases of Barbasol shaving cream were often consumed at both fossil fuel and nuclear generating stations. A technician would spread the shaving cream over the tubesheet and wait until it was sucked into the leaking tube, which repair personnel would then plug, thereby taking the tube out of service. Unfortunately, because tubes might also *appear* to be sucking in the cream without actually doing so, nonleaking tubes would sometimes become plugged unnecessarily. Further, not only was a nonleaking tube plugged, but prevailing standard operating procedure was to plug the surrounding tubes as well—a practice often termed 'insurance plugging'.

Other methods were equally crude. The technician might place a piece of plastic wrap over the tubesheet and look for an area where the plastic was being sucked in. Here again, the practice was to also plug nonleaking tubes surrounding the "suspect" leaker. Another technique was to enter the waterbox, partially close the manway, and then light up a cigarette and hold it in front of individual tubes to see if the smoke was "inhaled" by the tube. Simple hearing and seeing were also used. Some individuals believed that by placing an ear close to the tubesheet and/or by just looking at a tube, they could determine whether a tube was leaking or not. Unfortunately, out of millions of tubes that have been inspected and thousands that have leaked, very few tubes have been "heard" to leak.

All of these intuitive techniques had their shortcomings in reliability, accuracy, and cost-effectiveness. None provided any means for verifying that a suspected tube was in fact the leaking one without first putting the condenser back on-line and then checking the water and/or off-gas chemistry. None of these techniques was supported scientifically and they all relied on the "gut feeling" of the technician.

12.3 EARLY DEVELOPMENT OF LEAK DETECTION TECHNOLOGY: HELIUM

In the past, the question most often asked with regard to water *inleakage* was, With which waterbox was the leak associated? At most generating stations, the routine way of identifying the leaking waterbox was to drain each waterbox in turn and check the condensate chemistry to see if the problem had disappeared. To increase the accuracy and reliability of condenser tube leakage detection, a method was devised utilizing a mass spectrometer and helium as the tracer gas. The project, sponsored by EPRI, assumed that a tube leak could be detected by spraying helium into tubes and sampling the condenser off-gas. If helium were to be injected into the circulating water while the turbine was under load and the condenser was operating under vacuum, the gas would be drawn into the condenser through the leaking tube and thus be evacuated with the rest of the non-condensibles through the condenser air-ejection system, where the presence of helium could be detected.

Because this was a new application for mass spectrometry, there were problems with the increased background helium, difficulty with isolation of the leaking tube, and sometimes detection of leaks even where no helium had been sprayed. This led to an initial uncertainty whether tracer gas leak detection in condensers would prove to be effective. However, determination and perseverance triumphed, but even so, tracer gas leak detection came to be perceived as more of an art than a science.

The first successful application of helium to detect *water inleakage* used a nitrogen "kicker" to make sure that the helium traveled the full length of the tube. After one waterbox was drained, a plenum measuring 1 ft wide and 2 ft tall and approximately 1 in deep was placed on the tubesheet to cover a group of tubes. Helium was then sprayed down the tubes for approximately 10 seconds; and immediately afterward, an equal application of nitrogen was used to "push" the helium down the tube. The subsequent installation of air movers in the manways on the side opposite the plenum ensured that the entire length of the tube was covered by the helium, so that the nitrogen was no longer required. By August 1978, the use of helium as a tracer gas for tube leak detection had become standard practice at nuclear generating stations.

During that same year, the question arose whether helium mass spectrometry could be used to locate the source of condenser *air inleakage*, and tests were conducted again using a mass spectrometer and helium as the tracer gas. Initial problems with this method of air inleakage inspection included an increase in background helium, difficulty in determining the exact location of a leak (e.g., whether it was associated with the packing or with the flange joint), and the fact that helium, being lighter than air, tends to rise. In spite of initial ambiguities in data interpretation, technicians tended to climb the learning curve faster as they performed more inspections. Eventually, proper and intelligent interpretation of the data became routine.

As technicians' proficiency grew and the "art" of tracer gas leak detection became standard practice within the utility industry, generating stations were reducing the amount of air inleakage as well as promptly locating condenser tube leaks. Unfortunately, it was found that in situations where there were only small water leaks, leaks closer to the outlet end, or leaking plugs, testing with the on-line injection of helium frequently gave no indication of a leak. Using helium to discover the leakage source of dissolved oxygen was also unreliable. Thus, it became clear that a tracer gas with higher sensitivity was needed, since nondetection of helium was proving to be an uncertain indication of the absence of a leak in the waterbox.

12.4 SULFUR HEXAFLUORIDE (SF$_6$) AS A TRACER GAS

A more sensitive tracer gas had to be identified if the technique of tracer gas leak detection were to evolve. Simmonds and Lovelock [1976] had found in England that sulfur hexafluoride (SF$_6$) could be used very effectively as an airborne tracer in atmospheric research [Tsou 1996]. The utility industry in the United States was also exploring the path of plumes from smokestacks and cooling towers using the same tracer gas. The fundamental property of SF$_6$ is that it can be detected in very low concentrations—as low as 1 part per 10 billion (0.1 ppb)—compared with the lowest detectable concentration of helium of 1 part per million above background. It was later found that on-line injections utilizing SF$_6$ also allowed leaks as small as one gallon per day to be detected.

Sulfur hexafluoride, discovered in 1900, is a colorless, tasteless, and incombustible gas which is practically inert from a chemical and biological standpoint

[Ullman's 1988]. It does not react with water, caustic potash, or strong acids and can be heated to 500°C without decomposing. Two of its common uses within the utility industry are for arc suppression in high-voltage circuit breakers and the insulation of electric cables. SF_6 also has many other uses, such as for etching silicon in the semiconductor industry; increasing the wet strength of kraft paper and, in the magnesium industry, protecting molten magnesium from oxidation.

12.4.1 Analyzer to Detect SF_6 in Condenser Off-Gas

In the early 1980s, EPRI [1988] sponsored the development of an analyzer to detect the presence of SF_6 in condenser off-gas. Known as the Fluortracer®Analyzer, it was also based on mass spectrometer technology. Figure 12.1(a) is a general view of the SF_6 analyzer itself, while Figure 12.2 provides a schematic flow diagram of a portable analyzer, additional details internal to the analyzer being shown in Figure 12.3. For the analyzer to detect the presence of tracer gas in the received sample, it is important that it be free from both mois-

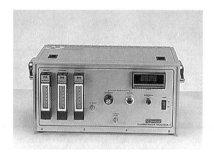

(a) SF_6 Fluorotracer™ Analyzer

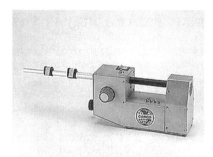

(b) SF_6 Dispenser SF_6 Pak

FIGURE 12.1. Photograph of (a) fluortracer analyzer and (b) portable SF_6 dispenser (Source Conco Systems, Inc.).

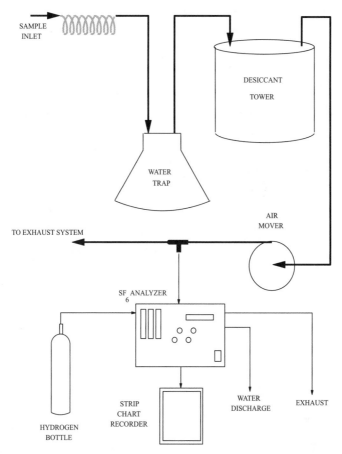

FIGURE 12.2. Flow schematic of the SF₆ sampling system and portable analyzer.

ture and free oxygen. Thus these diagrams show how the off-gas sample is first cooled, the moisture being removed in a water trap, then passed through a dessicant tower to remove any residual moisture, and is finally received by the analyzer. Then, to remove any oxygen contained in the sample from the off-gas system, hydrogen gas is introduced into the sample stream, entering with the sample into the catalytic reactor, where a chemical reaction occurs between the oxygen and hydrogen. The moisture so produced is removed by another water trap and dessicant tower. As shown in Figure 12.3, the dried sample gas is then pumped into the electron capture cell, where it passes between two electrodes and is ionized by a radioactive foil. Ionized nitrogen in the sample supports a current across the electrodes, the current level being reduced in proportion to the concentration of SF_6.

An analyzer for use with SF_6 as a tracer gas is commercially available for

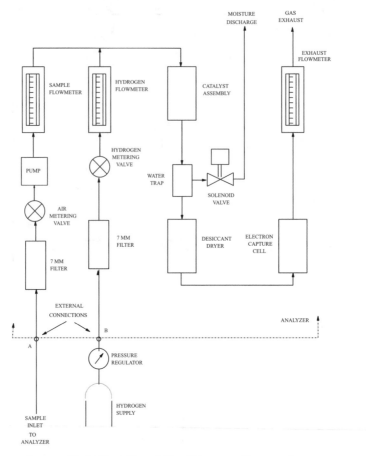

FIGURE 12.3. Details of the SF$_6$ sampling system.

use in both fossil fuel and nuclear generating stations. The analyzer is also provided with an SF$_6$ dispenser, shown in Figure 12.1(b) weighing approximately 8 pounds, the flow from which can be adjusted so as to obtain the SF$_6$ concentration needed to perform the leak detection task under current plant conditions.

12.4.2 First Successful Application of SF$_6$ Technology

The first on-line injection using SF$_6$ was conducted at a nuclear generating station and proved to be successful [Strauss 1992]. The on-line testing continued over a period of six months, and it was found that in the waterbox under test a tube leak was being detected with certainty whenever an SF$_6$ indication was seen on the strip chart recorder.

Again, the natural progression led to utilizing SF$_6$ for condenser air inleakage inspections as well. Before the use of helium, generating stations were running in excess of 50 ft³/min of air inleakage. With the introduction of helium as a tracer gas, the air inleakage at most stations was steadily brought down to much lower levels. Now stations would be running with less than 10 ft³/min of air inleakage. However, as the rate of air inleakage falls, it becomes more difficult to find leaks when only helium is used. In the search for a tracer gas with an improved sensitivity and reliability, SF₆ was found to overcome those difficulties encountered with helium.

12.5 STRIP CHART RECORDER VERSUS SPECTROMETER READOUT

Tracer gas leak detection involves a time delay between the injection of the gas and the response shown on the strip chart recorder connected to the mass spectrometer. It should be noted that, while the indicators mounted on the case of the mass spectrometer display the instantaneous values of the measurements, they are difficult to interpret without the addition of a trace from a strip chart recorder, a typical example of which is shown in Figure 12.4. The information available from the recorder chart not only provides a hard copy for future use but also tells technicians when they are getting close to a leak or have passed the leak, when they have located the leak, whether the gas is traveling to another leak, or whether the leak is closer to the outlet end. Whether a valve is leaking at the packing as opposed to the flange can also be determined.

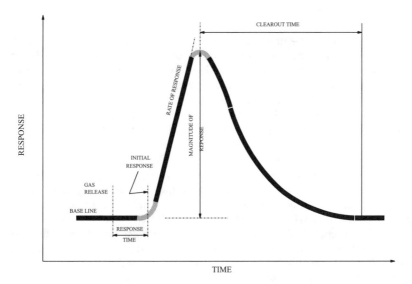

FIGURE 12.4. Chart recording of a typical leak response.

12.6   TUBE WATER LEAK DETECTION

The existence of a cooling-water leak from a tube is often first detected by an increase in the conductivity of the condensate. The chemistry department should be asked to verify the source of the contamination before plans can be made to locate the problem and rectify it.

It is most important that a test shot be taken *prior* to the formal tube leak inspection to verify that all of the equipment is working properly and that a proper sample is being received from the off-gas. The worst thing that can happen to a technician performing a tracer gas inspection is to finish the inspection without finding any leaks and then discover that no tracer gas was being sensed. The second reason for performing a test shot, important for both air inleakage and tube leakage inspections, is to determine the typical response time. It is almost impossible to find and isolate a tube leak efficiently without knowing the response time. Absent this knowledge, technicians may tend to chase every indication of leakage that they encounter. Thus, in addition to familiarity with the use of the detector, the two most important elements for a successful leak detection program are: (1) knowing the response time for the installation under inspection and (2) making effective use of the strip chart recorder. Technicians should also be discouraged from prejudging situations, since this can give rise to false data and result in delays in locating the leak.

SF_6 can be used whenever and wherever helium can be used, although the reverse is not true. A number of factors go into the decision of which tracer gas to select for a given application. Standard procedure is to use SF_6 as the tracer gas to determine where a waterbox is leaking. Helium has only a 50/50 chance of succeeding if the leakage is less than 100 gallons per day. The only alternative approach is to reduce power, drain the waterbox, and then look for a change in condensate chemistry. The following factors should be taken into consideration:

- *Size of leak.* If the chemistry shows a leak in excess of 50 gallons per day, then either SF_6 or helium can be utilized. If leakage is less than 50 gallons per day, then use of SF_6 should be the standard procedure.
- *Unit turbine power.* If the unit is running at greater than 20% turbine power, either tracer gas may be used. If the unit has no turbine power and the leak is so bad that the unit cannot be brought up to any turbine power level, standard procedure would dictate the use of helium.

When using pure SF_6 for on-line circulating-water injection, all gas fittings must be tight and connections to the regulator and to the circulating-water line must not be leaking. Most important, when connecting the hose to the injection point, quick disconnects must be utilized to ensure as little leakage as possible. The two most common causes of an increase in background tracer gas are carelessness when disconnecting the hose from the regulator and pressurizing the injection hose prior to connection to the circulating-water line injection point. Both are contrary to written procedures.

12.6.1 Condenser Tube Inspection

It is often helpful to *approximately* locate a leak before finding its exact location; a method for doing so that can be used as a last resort is to inject tracer gas onto the tubesheet while the waterbox is being drained. This is because, while a leaking tube row can be identified, other leaks in lower rows will not be clearly identified unless the leak(s) in the upper row have first been sealed.

Awareness of the approximate location of the leak (i.e. whether closer to the outlet end, or a leaking plug, or a waterbox seam) can help prevent error in finding the exact location of the leak. However, once the unit is down-powered and the waterbox is drained, the technicians must systematically check the entire tubesheet from top to bottom, since one large leak could be masking a smaller leak in a different location within the box. Furthermore, effectively locating a leak requires very discrete application of the tracer gas, and it is important to keep a record of every shot of tracer gas during the whole condenser tube leakage inspection process; otherwise, isolating that one leaking tube can become almost impossible.

A general setup for testing tubes for water leaks is shown in Figure 12.5. Experience has shown that the use of a plenum placed directly on the tubesheet is very helpful (see Figure 12.5). Typically, a 1 ft × 2 ft × 1 in plenum is used first. If leakage is indicated, then a 1 ft × 1 ft × 1 in plenum should be used, followed by a 4 in × 4 in × 1 in plenum, and ultimately, after a series of reductions, a "single-tube shooter." The plenum reduces the possibility that a tube will be missed or that tracer gas will be dispersed across the whole surface of the tubesheet. Finally, a wand-type mechanism should be utilized to spray the waterbox seams after the tubesheet inspection has been accomplished. Note that 99.9% of waterbox inspections begin on the inlet-side waterbox, spraying the tracer gas toward the outlet end.

Once it is determined that the strip chart recorder is functioning correctly and the sample response time is known, the tester typically starts inspection of a waterbox from the upper left corner of the tubesheet, working toward the right-

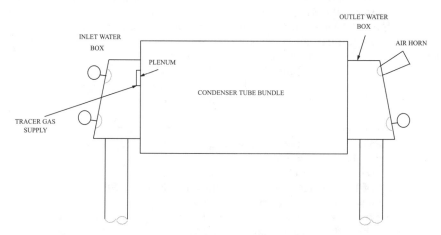

FIGURE 12.5. General setup for tube water leak test.

hand side of the box, and then drops down a row and works from right to left, and so on. This alternating sequence of left to right and right to left continues for the entire inspection. Normally, each shot of tracer gas lasts for 10 seconds. After waiting a further 2 seconds, the tester moves the plenum for the next shot. Unless the technician monitoring the detector sees an indication of leakage, the tester in the waterbox will continue to shoot without stopping.

There are instances where a leak appears to be indicated with the very first shot as well as with all subsequent shots. This should raise suspicions, and the first action should be to check the response time. Is it too long, or right on time? If the response time is too long, the reason may be one of the following problems:

- The leak is located closer to the outlet end. To determine if this is so, simply take the first 10-second shot, wait before shooting again, and monitor the strip chart recorder. If the recorded response time plus 15 seconds expires with no indication, repeat the same step until you duplicate the leakage indication. If the response time is right on target, obviously an area of leakage has been found and descending-size plenums must be used to isolate the leaking tube.
- The gas evacuated from the outlet waterbox is being directed by the blowers to an air leak location. To eliminate this possibility, simply spray the tracer gas into the suction side of the blower and monitor the strip chart recorder for an indication. If an indication appears, redirect the blower or add elephant trunking to direct the exhaust outside the building.

12.7   AIR INLEAKAGE DETECTION

Air inleakage can be inferred from an increase in the air concentration in the gases drawn off by the air-ejector system. It is often associated with an increase in condenser back pressure. There is always a minimum air inleakage which cannot be eliminated. Westinghouse used to recommend that air inleakage levels be held to 1 ft^3/min per 100 MW of generation capacity, but other minima are given in the ASME and HEI standards. An increase in the dissolved oxygen concentration in the condensate can also indicate that air is leaking into the suction of the condensate pumps below the condenser hot well.

As was true for water inleakage detection, SF$_6$ can be used whenever and wherever helium can be used, but the reverse is not true. Among the factors that go into the choice of tracer gas for a given situation are the following:

- *Air inleakage into the unit.* If the unit has more than 10 ft^3/min of air inleakage, either tracer gas may be used. If the inleakage is less than 10 ft^3/min, then SF$_6$ should be used.
- *Dissolved oxygen (DO).* The standard procedure for searching for the cause of DO leakage below the steam space is to use SF$_6$.
- *Unit turbine power.* If the unit is running at 20% or greater turbine power, either tracer gas may be used. If the unit has no turbine power and cannot be brought up to any level of turbine power, then helium should be selected.

- *Unit size.* Inspections of units of less than 50 MW capacity should always use helium.

Air inleakage inspections are best preceded by a test shot to establish the level of any background contamination and make sure the instrumentation is functioning correctly. It is recommended that all air inleakage inspections begin on the turbine deck, usually starting with the rupture disks. When on the mezzanine level, the tester should start spraying tracer gas at the condenser and work outward along the hood, the expansion joints, and other potential sources of air inleakage. A fan or blower installed so as to evacuate the outlet waterbox will assist in drawing the tracer gas down the tubes. Also, by evacuating tracer gas from the opposite end of the waterbox, the tracer background concentration will not rise to the point where detection of small differences becomes virtually impossible. (As a side benefit, the blower will increase workers' comfort.)

It is important to keep track of everything that is sprayed with the tracer gas. During a typical air inleakage inspection it is not uncommon to spray tracer gas on literally hundreds of suspected leakage paths within the condenser vacuum boundary. Thus, in order to isolate a leak as quickly as possible, it is important for technicians to know where they have been and what they have seen. If a large leak is found on the manway on, say, the west side of the turbine, an indication of this must be made on the strip chart recorder, because, when the technician goes to the west side of the condenser and sprays a suspected penetration into the condenser on the mezzanine level, the large leak could very well cause an erroneous indication of leakage there. Technicians can also waste a lot of inspection time searching for a leak that they have already found on the turbine deck. This is another reason why the typical response time must be known.

Once the inspection results are in, it is important to understand what can and cannot be done with them. A leak detection program is useful only if there is a follow-up repair program. Both SF_6 and helium detectors give readouts, one in millivolts, the other in an arbitrary scale. Plant personnel can determine a plan of action to repair the leaks after comparing millivolt or scale readouts. These are *relative* values, not calibrated in engineering units such as ft^3/min. Nor is it of any importance to have exact leakage values. Generating stations already know what their *total* air inleakage is; because of the margin of error due to all the variables of a condenser under vacuum, the quantifying of leaks would not add any information to what the plant personnel already know and would not, therefore, be cost-effective. What *is* useful is the exact location of each leak and its subsequent repair and retest.

12.8 PERFORMING TUBE LEAKAGE OR AIR INLEAKAGE INSPECTIONS WHEN TURBINE IS NOT UNDER POWER

When the turbine is not under power, it is very likely that the background concentration of the tracer gas will become so high that it eliminates any chance of isolating a leak. Both air inleakage and condenser tube leakage inspections require vapor flow to carry the tracer gas out of the condenser with the rest of the non-condensibles. If a sprayed tracer gas is sucked into the condenser, it will begin to accumulate, and the background concentration will rise and even saturate the detectors. There are, certainly, some occasions when, in an effort to bring a unit back up on-line as soon as possible, a station has no choice but to attempt a tracer gas inspection with the unit shut down, and many have been successful in doing so. However, a minimum 20% turbine power is recommended in order to perform a tracer gas inspection effectively.

12.9 USING TRACER GASES FOR THE INSPECTION OF OTHER SYSTEMS

In addition to condenser tube leakage and condenser air inleakage inspections, tracer gas leak detection is routinely performed on main-generator hydrogen cooling systems, and stator water systems. Applications in other industries have included testing of mine ventilation systems and searches for leaks from buried natural gas pipelines. Basically, if a system is under vacuum or can be pressurized, the use of a tracer gas is the most accurate, reliable, and time-effective method.

12.10 EDDY-CURRENT TESTING

Eddy-current testing (ET) is the technique employed for ascertaining whether condenser tubes have become pitted, corroded, or cracked; for determining the depth of such blemishes and their angular location; and for measuring their distance along the tube length. It is usually conducted to locate tube leaks, and to prevent future leaks, because they allow cooling water to contaminate the condensate. This contamination can upset condensate chemistry as well as cause possible corrosion problems elsewhere, especially in the boiler or nuclear reactor. The information obtained from ET, conducted at several points in time, can also help in scheduling maintenance and planning a tube replacement or plugging strategy, all with a view to extending the tube life of the condenser as well as avoiding unscheduled unit outages.

Eddy-current testing is a nondestructive test technique in which electrical currents are induced in the material being tested; the associated magnetic flux distribution within the material can then be observed. The eddy currents are generated by electromagnetic coils located within the test probe, and these are simultaneously monitored by measuring the electrical impedance of the probe. The material must be conductive, and, since electromagnetic induction is involved, there need be no direct contact with the sample.

The large number of potentially significant variables in ET is both a strength and weakness of the technique. Virtually everything that affects eddy-current flow, or otherwise influences probe impedance, must be taken into account, recognizing that important information can sometimes be masked by the effects of comparatively insignificant parameters. Thus, credible eddy-current testing requires a high level of operator training and awareness.

Before testing the tubes in a condenser, the ET system must first be calibrated using a sample of the same tube material. Probes are usually either of the bobbin or surface type, and for best results, the effective diameter of the probe should be close to that of the inside diameter of the tubes, due allowance being made for tube manufacturing tolerances. Clearly, when the tubes in the condenser itself are to be tested, they should have been brought to a clean state, to minimize the possible effects on the electromagnetic flux distribution caused by deposits.

Eddy-current testing can detect at least four kinds of damage in condenser tubes:

1. Corrosion pitting
2. Crevice corrosion
3. Fractures caused by tube vibration
4. Through-wall penetrations

In the first three of these, the depth of penetration is an important benchmark; it can lead to a decision to plug the tube as a precaution against future leaks. Identification of through-wall leaks will, of course, call for them to be plugged, immediately after the testing has been completed.

12.10.1   Eddy-Current Data Acquisition and Processing Practices

The tracer gas method of testing for water inleakage outlined in Section 12.6 can quickly locate the tube which is the source of a leak so that it can be fixed, but it is not a technique which can establish the overall condition of the condenser tubing and so contribute toward future informed maintenance planning. Eddy-current testing is able to provide this information. Unfortunately, the testing is often conducted only when tube leaks are occurring frequently, and are so severely affecting condensate chemistry that they cannot be ignored. In such situations, unless the problem is corrected, a forced outage of the unit will occur.

While such a use of eddy-current testing can provide a *snapshot* of the current state of the condenser, it does not provide an understanding of the *rate of growth* of pitting or corrosion, in order that steps can be taken early in the life of the condenser to mitigate their effects and so delay the time when complete retubing may become necessary. A better approach is to conduct routine eddy-current testing on at least a yearly basis, so that the progress of corrosion can be analyzed and the effects of any mitigation steps quantified. However, this data has to be archived and means have to be provided so that it can be accessed without difficulty and intelligently compared from year to year, the test results having been analyzed and even plotted. A properly planned program would also allow a comparison of the progress of corrosion on similar units at the same site.

A further consideration is whether 100% of the tubes should be tested at each inspection, or whether only subsets of tubes should be tested, once the principles to be used in their selection have first been defined.

12.10.2 Planning Eddy-Current Testing

Experience has shown that the greatest benefits from eddy-current data acquisition and subsequent processing can be obtained only through careful planning over a long time horizon. Clearly, the method of performing a planned eddy-current test should be thoroughly specified. In addition, analysis of the data on a year-to-year basis with the help of personal computers, and comparison of the results of tests taken over several years, should also be an important part of the overall project. How else can the results of each test be compared and sound judgments made regarding the effects of different maintenance methods or of the rate of tube deterioration?

Conley [1983] was an early proponent of systemwide corrosion monitoring. He stressed the need to provide advance notice of existing corrosion problems, feedback to operators regarding the adequacy of the corrections they had made, and determination of the component corrosion rate. Kaul and Willey [1990] and Singh et al. [1990] described computer databases which could compare data from one examination to another and create tube maps which could be displayed pictorially. But equally important ideas have largely been undiscussed: the figures of merit, and ways to plan the inspections so as to increase the confidence level in the data, especially when only a small percentage of the tubes were being tested. The following subsections suggest the set of principles which should be followed in planning the long-term eddy-current test project for a condenser or heat exchanger.

Establishing Procedures

Part of the planning process consists of writing of well-thought-through eddy-current test procedures for the whole set of annual or periodic tests, insisting that they are conducted in exactly the same way for each inspection, using the same calibration pieces and the same type of equipment for each test, and employing operators who have each been subjected to the same training and qualification requirements (ASME Boiler and Pressure Vessel Code, Section V, Appendix 8—ASNT SNT-TC-1A). In the United States, where the bidding process may result in hiring a different contractor for each inspection, the necessity for a common procedure will tend to ensure that eddy-current data from one inspection can be compared *with confidence* with the data from inspections conducted in earlier years. Without such formal procedures, and owner supervision to ensure that they are implemented, reliable data trending is virtually impossible.

It may be objected that such firm adherence to known procedures tends to inhibit technological advances. It is certainly true that they will be introduced more gradually. Clearly, the absolute values obtained from eddy-current testing are important in determining the damage mechanism and the appropriate corrective action to be taken now, e.g., which tubes to plug. But the ability to determine rate of change of corrosion is just as important with regard to determining when, or even whether, to retube.

Innovation can, however, take several forms. Holt et al. [1993] describe aural algorithms which they developed for eddy-current and other NDE (non-destructive examination) signals and which they used to alert the inspector when a significant blemish was detected in a tube. Besides displaying the electronic signal on the video display, aural information was presented to the operator through headphones, with the phase angle represented as the pitch of a musical tone and the magnitude of the depression represented by the loudness. This form of technological enhancement can be applied immediately, without casting doubt on the data acquired, either now or in the past. Other technological improvements, such as in basic measurement methods, may take longer to become common practice, since the data obtained from new measurement methods may not be directly comparable with that obtained from earlier examinations conducted using different measurement techniques.

Tube Map

The whole tube map should be defined at the outset and the tube numbering maintained throughout the life of the unit, and for all examinations. The configuration of this map should be well controlled, and changes should be made only in a very orderly fashion. Tubes which have been plugged should be clearly identified, and at each inspection, the plugs should be checked to ensure that they are still in place and leaktight.

Benchmark Data Set

It is important to have a benchmark ET data set for the condenser tubes, taken preferably before the unit is started up, and against which future changes can be compared. The eddy-current calibration tube should preferably be taken from the same manufacturing batch as the tubes installed in the unit itself. The calibration standards should be stored in a known place and used for each subsequent examination of the condenser or heat exchanger.

Cleaning Tubes Prior to Testing

Before conducting an eddy-current inspection, the tubes must be mechanically cleaned thoroughly; otherwise, misleading signals can be generated by larger deposits, debris, or other obstructions. Mechanical cleaning will also interrupt the pitting, while the water used to propel the cleaners through the tubes will also flush out any foreign material.

Data Disk Format

The eddy-current data tape or disk that is to be created during the examination should be capable of being transformed, with minimal manual intervention, into data which can be used by a computer program. The alternative of manually loading eddy-current data into a computer file is a tedious, error-prone process.

Data Comparisons and Trending

The program used to analyze the data need not be extremely sophisticated so that it becomes a major systems engineering project. For example, some users have simply written their own BASIC language program containing only about 2400 lines of BASIC code. The data for each inspection should be contained on its own disk so that it can be analyzed independently. In the comparing mode, the data from several inspections should be able to be read in, one set at a time; processed; and then compared. Since tube maps can be created for only one set of inspection data at a time, the plotting program can be separate from the analytical program, provided it is written to accept the same data file layout and format. The comparison of data sets, or trending, is made easier if 100% of the tubes are inspected every time. However, if only a subset of tubes is to be examined, this procedure requires more careful planning so as to ensure a high level of data continuity from one inspection to another.

Maintenance Practices

It is important to log all changes in maintenance practice as they occur, in order that changes in corrosion rate can perhaps be correlated with such procedural changes. For example, on one site, mechanical cleaning was identified as the cause of the reduced corrosion rate [Putman et al. 1994]. Changes in water treatment practice should also be logged and a summary of all of these changes provided to the inspector before the new eddy-current inspection is carried out.

Figures of Merit

The criteria for figures of merit are different for condensers and heat exchangers. In the case of heat exchangers, considerations of meeting the Pressure Vessel Code override questions of mere wall penetration. Some condenser users have chosen as a figure of merit the mean annual corrosion velocity and have used this to predict when a condenser should be retubed. Other figures of merit may be based on the square root of the sum of the mean of the squares of the corrosion depth. Conley [1983] suggested three possible criteria: equilibrium corrosion rate, time-weighted average equilibrium corrosion rate, and average corrosion rate. Clearly, as the number of plugged tubes increases, the reduced capacity of the condenser to remove heat from the vapor must also be taken into account. Thus, there should be some agreement on how corrosion figures of merit are to be defined.

Management Report

It is a common complaint that eddy-current test reports are so full of detailed data that it is difficult for management to appraise the results and make appropriate decisions. For this reason it is recommended that a summary, or executive report, be provided from which management can understand the significance of the results within the carefully structured long-term plan.

12.10.3   Testing Subsets of Tubes

Problems certainly arise when only subsets of tubes are tested. It is common practice during the early life of the unit to visually examine the condenser or heat exchanger and determine those tubes which seem to have the worst corrosion. The areas adjacent to these tubes are then eddy-current tested thoroughly and the data recorded. A given percentage of all other tubes may also be examined, evenly distributed over the remaining areas of the tubesheet. However, from inspection to inspection, while the *percentage* tested may not change, the *identities* of the tubes can be different.

This practice requires that the tubes being tested during any given inspection be carefully identified from the tubesheet map, in order that data from subsequent tests can be compared with confidence. All of this requires time in planning and care in execution. Conley [1983] suggested a tube assignment scheme as follows:

- Assign one half of the subset to peripheral tubes.
- Assign a random but large number to condensing zone tubes.
- Assign 25% of all air-removal tubes.
- With brass alloy tubes, assign a random but large number to the tubes just below the air removal section.

However, to improve the usefulness of data taken when testing subsets of tubes, some additional guidelines are proposed. Assuming that a total of 10% of the tubes are to be examined, it is suggested that these be allocated to three groups:

1. A control group that consists of 2% of the total number of tubes, this group being examined during *every* inspection. In this way, a level of continuity is maintained from test to test so that some comparisons to determine the progress in pitting or corrosion can be made with confidence. In other words, a growth curve analysis can be performed on this control set. This 2% should probably be distributed evenly throughout the tubesheet, or it can be assigned to bundles according to the principles of stratified sampling.
2. An additional 4% should be assigned to the control set so as to form clusters of three, in the case of offset tube spacing; or clusters of four, in the case of equal vertical and horizontal spacing.
3. The remaining 4% should be assigned in clusters of three or four throughout the remaining space, or centered on those areas which, from experience, tend to corrode the fastest.

The purpose of the clusters is that they form a *simplex* which can be used in a manner similar to the simplex used in various optimization strategies, especially those of the simplex self-directing type [Carpenter 1965]. Thus, after every inspection, it will be possible to select that adjacent tube which the present and past data shows is highly likely to exhibit a greater corrosion than those which have already been tested. The principle is to identify the most corroded pair of tubes in a cluster of three, and to pivot the triangle about this pair, considering them as the base (see also Figure 12.6). In this way, the next eddy-current inspection will be preplanned so that the set of tested tubes will always be moving toward the area of greatest corrosion, while still allowing data from two of the tubes in each cluster to be compared from test to test.

If such a procedure is begun from the initial startup of the unit, there is a high likelihood that, by the time corrosion has become serious, the worst-corrosion area will already be included in the subset being tested. The ability to compare data from test to test for most of the subset is a vital feature, allowing the progress of corrosion to be trended with a high level of confidence in the quality of the information.

12.10.4 Case Study: Yong Nam Thermal Power Plant

The Research Department of the Korean Power Company has been routinely conducting eddy-current tests on condenser tubes, almost on an annual basis. The data has been used not only to determine which tubes should be plugged but also the rate at which corrosion was developing in other tubes, and within the condenser as a whole. Figures of merit were developed to estimate remaining life and when retubing should be scheduled. The data was also used to measure the effect of regular mechanical cleaning on corrosion rate and resulted in postponement of a planned retubing.

The Plant

The Yong Nam Thermal Power Plant is located in the city of Ulsan on the coast of Korea. A city of 1 million people, Ulsan has a large concentration of oil refineries and petrochemical plants owned by major corporations. The power plant, designed by AEG in Germany and commissioned at the end of 1970, contains two 200-MW reheat units, the boilers of which are fired with Bunker C fuel oil. Main steam conditions are 180 kg/cm^2 (2560 psig) and 530°C (986°F).

Many industrial plants upstream of the power plant discharge their sewage streams into the Taewha River, which runs through the center of the city and is also the source of cooling water for the plant before it enters the sea. Table 12.1 gives the water analysis over several years and shows that the water quality is poor. The chemical analysis, chemical oxygen demand (COD), and dissolved oxygen (DO) all violated the appropriate standard. The ammonia (NH_4) is especially troublesome, since the average value of 0.64 is in excess of the specified limit of 0.2 [KEPCO 1992, Kim 1993].

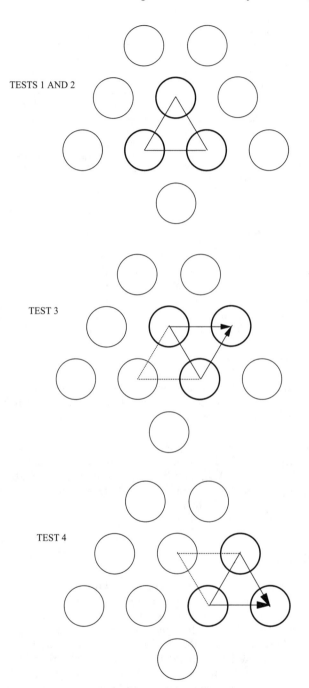

TESTS 1 AND 2

TEST 3

TEST 4

FIGURE 12.6. Search for the maximum-corrosion tube cluster.

Table 12.1. Cooling-Water Analysis

Attribute	Fouling Standard	Year				
		1989	1990	1991	1992	Average
pH	<7.5	7.95	7.85	7.70	7.70	7.80
NH$_4$	<0.2	0.77	0.44	0.74	0.61	0.64
COD	>4.0	3.88	2.95	3.03	2.63	3.12
DO	>6.0	5.72	5.48	6.10	6.47	5.94
Sulfide	Trace	Trace	Trace	Trace	Trace	Trace
Turbidity		9.05	11.23	21.38	22.10	15.94

Condenser Design

The unit has two-pass condensers which originally had 18,460 tubes with an outside diameter of 23 mm (0.91 in), made from aluminum brass (DIN SOMS78F40), and with a wall thickness of 1 mm (approximately equivalent to 19 BWG). The surface area is 10,000 m^2 (107,639 ft^2), and the condenser design back pressure was 722 mm-Hg (28.42 in-Hg). There are four neoprene-lined cast iron waterboxes, these and their tubesheets being arranged so that they form two 2-pass condensers.

The condensers were equipped with a Taprogge sponge ball tube cleaning system which uses 800 balls twice a week in the summer season but only once a week in the winter. In addition, a debris filter was provided at the cooling-water inlet. A hydrochloric acid generator was used for cooling-water treatment, while corrosion protection was provided by an electrical positive pole system in the waterbox.

Two cooling-water pumps were provided for the condenser, each with an individual flow capacity of 15,500 m^3/h (68,269 GPM) and a discharge head of 10 meters w.g. (32.81 ft w.g.). When both pumps were in operation, the flow rate for each pump dropped to 13,000 m^3/h (57,257 GPM), and they then had a combined head of 13 meters w.g. (42.65 ft w.g.).

Tube Failures

In February 1979, 9230 tubes were replaced with AP-bronze (JIS) tubes produced by Sumitomo in Japan. The composition of this tube material was Cu, 91%; Sn, 8%; Al, 1%; Si, 1%. It was much more resistant to seawater corrosion. The remaining 9230 tubes on this condenser were replaced by the same material in August 1980. However, due to repeated occurrences of severe tube leaks, the need for another retubing in 1991 was considered a possibility. Since 1980, the condition of both tubes and waterbox had been carefully monitored [KEPCO 1992]. Not only had there been a significant reduction in the number of tubes which had to be plugged, but there were several years in which no leaks were experienced at all.

Table 12.2. Tube Failure Incidents

Year	Plugged	Leaks
1980	1	
1981	2	1
1982		
1983	7	5
1984	7	
1985		
1986	34	
1987		
1988	9	
1989	NA	NA
1990	2	2
1991	8	4
Total	69	13

The failure incidence of the AP-bronze tubes since they were finally installed in 1980 is outlined in Table 12.2. Note that the plant was subjected to a long-term outage from August 1984 through October 1986, which accounted for the need to plug so many tubes in 1986. However, no new tube leaks were detected in 1987. The absence of any new tube incidents in 1989 is because no inspections occurred that year.

Water Treatment Practice and Corrosion Protection

It had been the practice to inject 1 ppm of $FeSO_4$ for one hour per day, while chlorine injection had been continuous at a rate of 0.3 to 0.4 ppm delivered to the inlet of the circulating-water pumps. The corrosion protection system was of the electric positive pole type, consisting of four lead-silver (Pb-Ag) anodes installed in each waterbox, a zinc reference cell, and an auto/current controller. The electric potential was checked routinely and was held close to the design value of 150 mV.

Mechanical Cleaning

In September 1991 the condenser was mechanically cleaned for the first time in 11 years by three passes of Conco metal tube cleaners. The 11 years of deposit accumulation, corrosion product growth, and underdeposit pitting required three passes in order to return the tubes to their design condition. However, in June 1992, only one pass of a Conco cleaner was needed to return them to their original condition. Subsequent analysis of the removed deposit by the KEPCO

Technical Research Institute showed an absence of shellfish within the tubes and tubesheet, but also showed the existence of thick deposits of $FeSO_4$ due to over-injection. The density of the removed deposit was found to be 132 g/tube, or 0.0253 g/cm^2, and its composition:

Cu as CuO	50.75
Cr as Cr2O3	0.24
Fe as Fe3O4	19.47
Na as Na2O3	0.21
SiO2	8.12
Zn as ZnO	0.22
Al as Al2O3	2.18
Ni as NiO	0.14
Mg as MgO	1.65
Mn as MnO	0.11
Ca as CaO	1.04
P as P2O5	0.00
S as SO3	0.34
Loss in weight	6.72
Insoluble	8.80

Sample tubes were examined both before and after mechanical cleaning. Prior to cleaning, a well-formed and thick protective sulfate coating was seen on the tube surface, and a new coating was seen to cover those areas which had become exposed. Some pitting corrosion was also observed.

Thin protective sulfate coatings provide good corrosion protection to copper alloy tubes. Unfortunately, they could become thick enough to reduce the tube heat transfer coefficient significantly [Conco 1985]. Since underdeposit pitting continues toward the tube outer wall, the pit must be cut open and flushed out to retard the pitting action [Saxon 1990].

Some pitting corrosion was observed. Closer examination of the pitted areas shows that the pitting was deposit-induced crevice-corrosion pitting. The deposit chemical analysis indicated that the primary foulants were iron oxide and silt. Some of the iron oxide clearly arose from the ferrous sulfate dosing.

After the sample was acid-cleaned in the laboratory, partial and slight pitting damage could be observed. The surface formation was checked under a magnification of both 200 times and 800 times without providing any further information.

Eddy-Current Inspections and Corrosion Analysis

The monitoring of corrosion was performed using an electric method as well as by means of periodic eddy-current testing. Seawater analysis was also monitored over the period. The eddy-current test equipment used by KEPCO was a Model ND-382 three-way eddy-current test system manufactured by Nippon Danso Co., Ltd., of Japan and utilized three different types of wave so as to produce a three-dimensional evaluation of the tube condition. The three techniques employed were:

Table 12.3. Analysis of Eddy-Current Inspections

	Number of Tubes Affected					
Corrosion Type	1987	1988	1989	1990	1991	1992
Even	1143	2633	NI	3041	3797	3628
Partial	1543	2643	NI	2934	4014	4152
Inlet area	55	96	NI	94	89	108
Plugged	41	59	NI	61	69	92
Totals	2782	5431	NI	6130	7969	7980
Percent	14.8%	21.9%	NI	32.9%	43.1%	43.2%

NI = no inspection.

- Absolute method
- Differential method
- Solution up method

The solution up method directly presents the percentage wall thickness in the corrosion area.

With this equipment, eddy-current inspections were conducted routinely, and Table 12.3 summarizes the results for the condenser as a whole. The total number of tubes affected by corrosion, or which had been plugged, increased by only 11 between 1991 and 1992, although the distribution among the different types of corrosion effect also changed. Again, note that no inspections were conducted during 1989. The accumulated number of tubes affected by each type of corrosion is shown in this table, while in Figure 12.6 are plotted the accumulated totals as well as the corrosion rate.

As shown in Table 12.3, it is noteworthy that for the tubes under survey, while the number of pitted tubes increased from 14.8% to 43.1% between 1987 and 1991, the increase in pitted tubes between 1991 and 1992 was only on the order of 0.1% and appears to be approaching an asymptotic state.

Table 12.4 shows the number of tubes with measured corrosion depressions of various depths. It will be seen that very few tubes (i.e., <10) show depressions which are more than 50% of the wall thickness. This is a good indication of the effectiveness of the KEPCO corrosion control program. Since annual use of the metal cleaners has been shown to double the tube life of condensers [Hovland 1988, 1990], KEPCO continued to monitor this closely. Finally, Table 12.5 shows the corrosion rate expressed in mm/year, and calculated from data obtained by KEPCO from the routine inspection of its tubes.

Commentary on KEPCO's Experience

Condenser inspections have shown no particular corrosion problems associated with either ferrous oxide injection, the use of sponge balls for on-line cleaning,

Table 12.4. Inspection by the Solution Up Method

Solution	Number of Tubes		
	1990	1991	1992
≤20%	5546	7080	6972
21–30%	455	703	734
31–40%	56	99	148
41–50%	5	15	25
>50%	7	3	9
Plugged	61	69	92
Totals	6130	7969	7980

the operation of the electric potential corrosion control system, or the use of metal cleaners for off-line cleaning. The principal corrosion damage was due to partial pitting. The substitution of AP-bronze tubes, which are more resistant to seawater corrosion, has resulted in a much lower corrosion rate. The significant reduction in corrosion rate which KEPCO experienced was attributed to an annual mechanical cleaning using metal cleaners to augment the chemical treatment and on-line ball cleaning procedures already being practiced.

Originally, it was thought that condenser 2 would need to be retubed in March 1991 because of corrosion. However, accepting the present annual average corrosion rate of 0.12 mm/year, it was anticipated that the present condenser tubes would remain in service through 1996, when the total amount of corrosion was to have become 0.62 mm. This has proved to be the case and the cost savings were a significant benefit, especially when added to the improved heat rate and unit availability.

12.11 PLUGGING TUBES TO REDUCE LEAKS

There are occasions when a tube wall becomes damaged by foreign objects or perforated as a result of the effects of vibration and/or corrosion. The resulting damage allows cooling water to leak into the shell side of the condenser and

Table 12.5. Average Annual Corrosion Rate

Year	Corrosion speed, mm/yr
1988–89	0.14
1990	0.14
1991	0.12

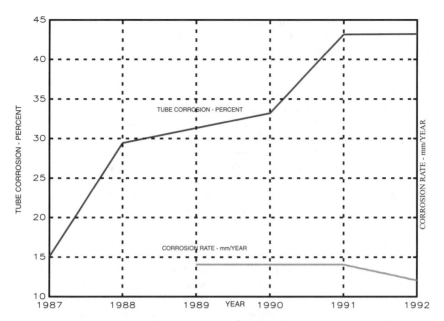

FIGURE 12.7. Corrosion progress for the Yong Nam example.

contaminate the condensate, either chemically or through its dissolved oxygen concentration. If the leakage is not rectified promptly, damage to steam generator tubes and/or turbine blading can result. As was described earlier in this chapter, leaking tubes can be identified using leak detection methods, as well as by eddy-current testing of some or all tubes in the heat exchanger. Once the tubes which have developed leaks are identified, they can be sealed off by plugging both ends, usually with special plugs made to some patented design.

The simplest plugs are tapered or conical in shape, made from a fiber material which is able to withstand temperatures of up to 230°F, and are easily installed when making emergency or temporary repairs (See Figure 12.8(a)). These plugs are dimensioned according to the size of the tube into which they are to be installed, and they expand when wet, effectively sealing off the tube. The same simple one-piece tapered design can also be made from a metal which is galvanically compatible with the tube material and expands when the temperature rises (See Figure 12.8(b)). Another form of tube plug consists of a metal collar inserted into the tube end, a tapered pin then being driven into the collar to provide a positive seal along the whole length of the plug (See Figure 12.8(c)). These plugs also expand with a rise in temperature, and where tubes have been extracted for sampling purposes, these plugs can also be used for closing the corresponding holes in tube sheets. Able to withstand high temperatures, plugs of this design can be used in heat exchangers in the utility or process industries, in addition to plugging the tubes in condensers.

All these types of plugs can be removed if desired, although some may have to be drilled out in difficult cases. However, for either temporary or permanent plugging of condenser or heat exchanger tubes, an expanding-cylinder plug configuration is also available. One such design, shown in Figure 12.8(d), consists of two thick flexible elastomeric sleeves mounted on a nut-and-bolt assembly, with one brass washer separating the sleeves and additional washers provided at the outer ends. As the nut is tightened, the sleeve cylinder expands and seals off the tube end. In one design the washer under the nut is large enough to fit snugly against the face of the tubesheet, thus preventing shell-side vacuum from subsequently drawing the plug down into the tube. On other plug designs the washers are smaller and allow the plug to be placed at a desired location within some distance from the end of the tube. All these plugs are suitable for use at maximum pressures ranging between 240 and 500 psig. Plug designs are also available with the same features but having only a single elastomeric sleeve or cylinder.

A slightly complicated but more secure design of plug is also shown in Figure 12.8(e) and consists of only one flexible elastomeric cylinder but is also provided with expanding metal grips mounted on the bolt. As the nut is tightened, the grip presses against the inside of the tube at the same time as the elastomeric cylinder is expanding, ensuring that the assembly is held securely in position by the grips, the expanded elastomeric cylinder providing the seal. Plugs made with this construction can withstand pressures in excess of 1000 psig and are also suitable for use on tubesheets which have been coated with epoxy or other material. All of these plugs designed with elastomeric cylinders can be removed later when convenient, so that an engineer can conduct a more complete investigation of the cause of tube failure without being under pressure.

Other plug designs are available, such as explosive plugs which require expert handling. For a review of a wider range of plug designs and their usage, see Chapter 10 of Andreone and Yokell [1998].

While plugging a few tubes will not significantly affect the performance of a condenser or heat exchanger, plugging too many tubes can reduce heat transfer capacity. The effect on water film heat transfer coefficient can be estimated from the following equation, where N indicates the number of tubes involved and h is the single-tube water-side heat transfer coefficient:

$$h_{\text{plugged}} = h_{\text{new}} \left(\frac{N_{\text{new}}}{N_{\text{new}} - N_{\text{plugged}}} \right)^{0.8} \tag{12.1}$$

Note that the film thermal resistance corresponding to h_{new} is only about 50% of the total tube thermal resistance. Clearly, there is some limit on how many tubes can be plugged without seriously limiting condenser duty, and the higher the design cleanliness factor, the fewer tubes can be plugged without affecting performance.

It should also be remembered that reducing the flow area by plugging tubes will increase both water velocity and condenser pressure drop. If ΔP represents tube pressure drop, then

$$\Delta P_{\text{plugged}} = \Delta P_{\text{new}} \left(\frac{N_{\text{new}}}{N_{\text{new}} - N_{\text{plugged}}} \right)^{1.8-2.0} \tag{12.2}$$

(a) Fiber Plug

(b) Brass Pin Plug

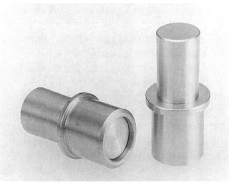

(c) Pin and Collar Plug

(d) Expanding Rubber Plug

(e) Expanding "High Confidence" Plug

FIGURE 12.8. Photos of tube plugs: (a) fiber plug, (b) brass pin plug, (c) pin and collar plug, (d) expanding rubber plug, (e) expanding "high confidence" plug (Source Conco Systems, Inc.).

Thus, the pump characteristic curves may play a part in determining the limit on the number of tubes that can be plugged. Further, while the higher water velocities will tend to increase the tube heat transfer coefficient, they will also increase the erosion/corrosion effects on the tube inner surfaces, providing another limit on how far tube plugging can be taken.

12.11.1 Installation Procedure for Plugs with Grips

When a tube is to be plugged, it is important that plugs be installed in both ends. Otherwise, if the tube has become perforated, plugging only one end will not prevent cooling water from entering through the unplugged end of the tube and continuing to contaminate the condensate. Further, stagnant water will remain within a tube plugged only at one end, and the corrosion rate will rise.

To install a plug, the first step is to ensure that the tube is drained and that the ends are dry. The nut provided on the assembled plug should now be tightened so that it is just finger-tight. The seal cylinder end of the plug should then be inserted into the tube, the grips being located as far as possible within the tubesheet and clear of any rolling transitions. The grips must engage the tube at a point where they have the tubesheet as a backing. Using a screwdriver to hold the bolt and a box wrench to turn the nut, tighten the nut until the screwdriver is no longer needed. Then, using a snap-on torque wrench having a 30 to 200 in-lb range and fitted with a deep well socket, torque the nut to 50 in-lb. Some 15 minutes after completing the plugging all those tubes which are to be taken out of service, retorque all plugs to 50 in-lb, or more if needed to conform to site-specific requirements. Torque ranges of between 50 and 100 in-lb should satisfy all applications.

12.11.2 Tests to Determine Plug Pressure Rating

It is important that plugs remain in place under varying pressures, temperatures, and vibration levels. If a plug should fall out of the tube, it is as though a new leak had developed, and high costs can be incurred by a new outage, or a reduced load operation, in order to rectify the situation. Thus, the performance of a selected design of plug should be evaluated with regard to both its blowout pressure and operational continuity under vibration conditions. A report by New England Power Service [1989] indicated that plug failure may sometimes be caused by flow-induced vibration.

Saxon et al. [1989] outlined a method used to evaluate the pressure rating for different types of plug. The objective was to establish the pressure at which blowout or leakage occurred and to understand how vibration might affect this pressure. The rig consisted of a frame mounted on a vibrating base having a range of 0 to 10 G's. A 36-in length of condenser tubing was mounted in the rig, the tube being provided with a heating element so that the tube temperature could be controlled. The ends of the tube were also rolled into thick aluminum blocks, used to simulate a tubesheet. The plug to be tested was inserted in one end of the tube, and a source of water at an adjustable pressure (230 to 1000 psig) was connected to the other end.

Prior to testing, the tube and aluminum blocks were heated to 112°F and held at that temperature for at least 30 minutes, to simulate the conditions under a vacuum of 2.5 in-Hg. Four tests were conducted for each plug, the pressure being increased in 25-psi steps until the corresponding blowout or leakage pressure was found:

1. Temperature held at 70°F—no vibration
2. Temperature held at 112°F—no vibration
3. Vibration set to 10 G's
4. Temperature held at 112°F—vibration set to 10 G's

The conclusions reached were that the blowout pressure was greatly affected by plug design. Increasing the tube temperature raised the blowout pressure by 5 to 15 psi, but short-term vibration reduced the blowout pressure. It is probable that long-term vibration would reduce performance even further.

Chapter 13

PERFORMANCE MONITORING OF POWER-PLANT HEAT EXCHANGERS

The condition of the various types of heat exchangers installed in power plants is important to the operating efficiency of the unit. However, the analysis of shell-and-tube type heat exchanger performance differs in several significant ways from that of condensers. First, the variations of temperature throughout a *condenser* are comparatively small. The exhaust vapor is at a pressure of usually not more than 5 in-Hg, and the vapor temperature throughout the vessel is sensibly the corresponding saturation temperature. Similarly, the water temperature rise is seldom more than 20°F, so that bulk water temperatures are relatively close to the inlet and outlet temperatures. For these reasons, the tube metal temperature is relatively uniform throughout the length of the tube and the variations do not affect the value chosen for metal thermal conductivity. Last, a condenser considered as a heat exchanger is usually of the single-pass cross-flow type. Even when it is of the two-pass configuration, the two passes may be considered as forming a continuous tube, and the direction of water flow is not important in the calculation of log mean temperature difference.

By contrast, power-plant *shell-and-tube type heat exchangers* are almost always of two-pass construction and generally operate at much higher temperatures and pressures than a condenser. The temperature variations are also much greater, affecting the value chosen for the metal thermal conductivity and thermal stress. Furthermore, the heat exchangers' construction is such that the vapor and water flows must be considered as having both parallel and counterflow relationships. This changes the form of the equation for log mean temperature difference and also requires applying a correction factor. The impact of these differences on the techniques for performance monitoring will be discussed in detail in this chapter.

The heat exchangers which are included as a part of the regenerative feedheating system on every turbogenerator unit, make an important contribution to the efficiency of the Rankine Cycle for both *fossil fuel and nuclear* plants. Figure 13.1 is a schematic diagram of a typical feedheating system. The condensate leaves the condenser at a temperature of about 100°F, and the feedheating system causes the water temperature to be raised to, typically, between 400 and 500°F before the feedwater enters the boiler, thus reducing the amount of fuel required to bring the steam to the design pressure and temperature. Figure 13.1 also shows steam being extracted from appropriate stages of the turbine to heat the water.

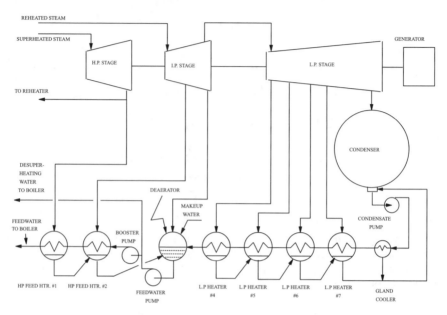

FIGURE 13.1. Schematic diagram of a typical regenerative feedheating system.

The higher the temperature of the water entering a given heater, the higher the pressure (i.e., turbine stage) at which the steam to that heater is extracted. This extraction is uncontrolled but attains an equilibrium condition when the product of steam flow and the latent heat in the steam is equal to the product of water flow and its enthalpy rise. It should be observed that, since the steam used for preheating is extracted from turbine stages, it also cogenerates power.

Although feedheaters are usually of the shell-and-tube type with two passes, the details can vary. Figure 13.2 illustrates a typical horizontal heat exchanger with straight tubes attached at one end to a tubesheet mounted adjacent to the fixed inlet head, the other end of the tubes being attached to a floating head. The fixed head is also provided with a baffle which separates the inlet and outlet water chambers within this head. Figure 13.3 shows a similar construction, but the tube bundle now consists of a set of U-tubes attached to the tube plate adjacent to the waterbox; this design dispenses with the need for a floating head. In both of these horizontal configurations, level controllers are normally provided to maintain an adequate condensate level within the vessel. Still other feedheaters are mounted vertically as shown in Figure 13.4. These can be of the straight-tube or U-tube type and are also provided with a condensate level controller.

At full load, all of these feedheaters should be operating close to their design conditions. However, under part-load conditions, both the feedwater and extraction-steam flow rates are correspondingly reduced, and the expected heat transfer coefficient under such conditions, even for a clean heat exchanger, becomes difficult to predict.

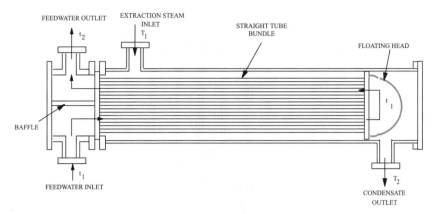

FIGURE 13.2. Schematic diagram of a double-pass horizontal feedwater heater, with straight tube bundle.

In *nuclear plants* there is another set of heat exchangers associated with the service-water system. Since these are usually part of the emergency system for the unit, normally they are either shut off or there is only a small flow of water through them. As a result, they are rather prone to fouling and corrosion problems. Clearly, the service-water system should function correctly when it is called into action in an emergency, but, because of their normally being in a standby state or shut off, these heat exchangers' performance or condition is difficult to monitor. The challenge is to be certain that they will perform correctly when called upon to do so. Furthermore, since the design data sheets for this type of heat exchanger state only the full-load conditions, it is again not easy to predict what the overall heat transfer coefficient should be under any lower load condition.

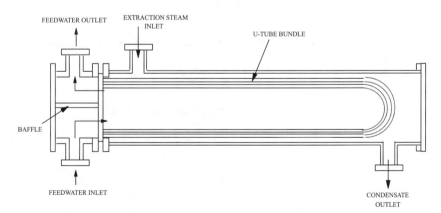

FIGURE 13.3. Feedwater heater similar to that of Figure 13.2, with U-tube bundle.

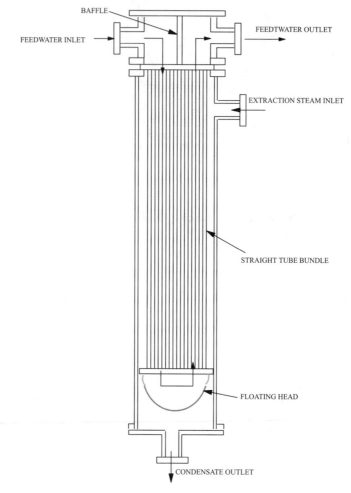

FIGURE 13.4. Schematic diagram of a double-pass vertical feedwater heater.

13.1 REGENERATIVE FEEDHEATING SYSTEM

The system shown in Figure 13.1 includes a turbogenerator with high-pressure, intermediate-pressure, and low-pressure stages, the last of them exhausting into the condenser. The condensate level in the hot well determines the rate at which condensate is drawn from the condenser by the condensate pumps, which pass it through a turbine gland cooler. From there it passes through up to four low-pressure heating stages, the heated condensate discharging from stage 4 into the deaerator. All except the deaerator are noncontact-type heat exchangers, whereas the deaerator is of the direct contact type, the pressure being that of the steam extracted from the outlet of the IP stage. As its name implies, the deaerator is the vessel within which any residual non-condensibles accumulate and from which they are vented from the system. (The condenser air removal system removes those non-condensibles that have accumulated in the condenser). The deaerator is also the vessel to which the boiler makeup water supply is usually connected.

The pressure in the system between the condensate pumps and the deaerator is on the order of 400 psia, while the extraction pressures range between 5 psia (heater 7) and up to 150 psia in the deaerator. Unfortunately, not all of the steam drawn from the turbine extraction points is superheated; thus, there can be uncertainties in the heat exchanger performance calculations. Also, while heaters 6 and 7 are shown as single vessels, these heating stages can consist in practice of two vessels connected in parallel. Low-pressure feedwater pumps usually draw the boiler feedwater from the deaerator. A high-pressure booster pump is also usually included, raising the feedwater pressure up to perhaps 3000 psia, before passing the water through the high-pressure heaters and into the boiler.

Note that the condensate drains from many of the feedheaters are connected back to whatever previous feedheater in the train is operating at a lower pressure. As the higher-pressure drains enter this lower-pressure heater, the liquid flashes off, the vapor helping to preheat the feedwater passing through the vessel before it condenses, while the liquid joins the condensate which has already collected in the vessel.

13.2 HEAT EXCHANGER MASS/HEAT BALANCE

Care must be taken when attempting to monitor the performance of a heat exchanger in the train shown in Figure 13.1, since full information is seldom available from the instrumentation and some mass flow calculations must be performed. This is particularly true for the extraction-steam flows, since they are seldom measured; while there are usually only two water flow measurements in the whole system, namely, the condensate flow at the discharge of the condensate pumps, and the feedwater flow to the boiler. Since the latter is the basis for heat rate calculations, it is usually a very accurate flow measurement.

Some mass flow calculation complications are also introduced by the cascading of the drains toward lower-pressure vessels. For instance, it is possible to calculate the steam flow to heater 1 from the enthalpy of the extraction steam and the enthalpy rise in the feedwater. But the extraction-steam flow to heater 1 is also the drain flow from heater 1, which is cascaded to heater 2. Thus, the steam flow to heater 1 must be calculated before performing any calculations for heater 2, and so on. Further, the drains from heater 1 are at a pressure higher than that in heater 2, so that the drains will flash as they enter the lower-pressure heater. Since the flashing vapor will be condensed, its latent heat must be added to the latent heat in the extraction-steam flow drawn by heater 2, in order that this extraction flow can be calculated with accuracy.

For these reasons, the flow calculations must commence with those for heater 1 and then be performed sequentially for each heater in turn along the train.

Heater 1

The pressure and temperature of the extraction steam will be known, so that its enthalpy (H_{stm1}) can be computed from the steam tables together with the saturated liquid enthalpy corresponding to the pressure (H_{liq1}). The feedwater flow (W_{fw}) is also known, together with the feedwater pressure and inlet and outlet water temperatures across this heater. Using the compressed water tables included in ASME (1993) to establish the enthalpy of the water at the inlet to H_{in1}, and that at the outlet from H_{out1}, enables the heat flux to be calculated. From this data, the *extraction-steam flow* W_{stm1} can be computed from:

$$W_{stm1} = W_{fw}\left(\frac{H_{out1} - H_{in1}}{H_{stm1} - H_{liq1}}\right) \tag{13.1}$$

Heater 2

The feedwater flow through heater 2 is usually the same as through heater 1, although, if water is taken off to supply boiler desuperheaters, its flow should be known and care must be taken to allow for this in the calculations. The pressure and temperature of the extraction steam to heater 2 will have their own values, and the steam tables can again be used to obtain the corresponding values of H_{stm2} and H_{liq2}. The inlet and outlet enthalpies H_{in2} and H_{out2} of the feedwater passing through heater 2 can be obtained similarly.

The flow of the drains from heater 1 will be the same as the extraction-steam flow W_{stm1} calculated for heater 1, but its enthalpy will be that of saturated liquid at the pressure of heater 1 (H_{liq1}). Meanwhile, the enthalpy of the saturated liquid in heater 2 (H_{liq2}) will be that corresponding to the pressure in vessel 2 and less than H_{liq1}. Thus, in addition to the latent heat to be removed from the extraction flow, there is an additional amount of latent heat in the flash-off from the drains from heater 1 to be removed, and this will tend to reduce the extraction flow to heater 2 below that which would otherwise be required. Thus, for heater 2:

$$W_{stm2}(H_{stm2} - H_{liq2}) + W_{stm1}(H_{liq1} - H_{liq2}) = W_{fw}(H_{out2} - H_{in2})$$

or

$$W_{stm2} = \frac{W_{fw}(H_{out2} - H_{in2}) - W_{stm1}(H_{liq1} - H_{liq2})}{H_{stm2} - H_{liq2}} \tag{13.2}$$

Heaters 4 and 5

Heaters 4 and 5 are handled in similar fashion to heaters 1 and 2, but the water flow is now the flow from the condensate pumps. This is usually lower than the feedwater flow rate by approximately the makeup water flow rate.

Heaters 6 and 7

Heaters 6 and 7 are also handled a similar fashion to heaters 1 and 2, with the water flow again the flow from the condensate pumps. However, if these stages have two heaters arranged in parallel, it should be assumed that half of the total condensate flow will be passing through each and the flows used in the mass flow calculations will have to be adjusted accordingly.

Heat and Mass Balances as a Set of Simultaneous Equations

It will be clear that the pair of heat and mass balance equations for each heat exchanger can be formulated as an interconnected set of simultaneous equations which can be solved in the conventional manner. All steam and water enthalpies can be calculated by presenting the appropriate set of pressures and temperatures to the ASME (1993) Tables for saturated and superheated steam and for compressed water and substituting these values in heat balance equations for the corresponding heat exchanger. The flow rate of the feedwater, high pressure desuperheating water, and either makeup water flow or condensate flow must also be known. This formulation of the problem has been demonstrated by Putman and Harpster (2000) and allows the effects of changes in selected parameters on the cycle as a whole to be studied.

13.3   EFFECTIVE HEAT TRANSFER COEFFICIENT Ueff

Reference has already been made to the parallel/counterflow steam/water flow relationships inherent in the design of these heat exchangers, and the calculation of log mean temperature difference must reflect this. Let:

$$T_1 = \text{temperature of extraction steam}$$
$$T_2 = \text{condensate temperature}$$
$$t_1 = \text{feedwater inlet temperature}$$
$$t_2 = \text{feedwater outlet temperature}$$

Then the log mean temperature difference for any heater can be obtained from:

$$\text{LMTD} = \frac{(T_1 - t_2) - (T_2 - t_1)}{\log\left(\dfrac{T_1 - t_2}{T_2 - t_1}\right)} \tag{13.3}$$

The heat transfer in heat exchangers can be described by a modified form of the Fourier equation (5.2), where A is the surface area of the tubes, CF is cleanliness factor, and F (to be discussed shortly) is the correction factor for log mean temperature difference:

$$Q = \frac{CF \times F U_{eff} A[(T_1 - t_2) - (T_2 - t_1)]}{\ln\left(\dfrac{T_1 - T_2}{T_2 - T_1}\right)} \qquad (13.4)$$

13.3.1 Design Cleanliness Factor

Section 2.3.4 contained a discussion of the meaning of the term *cleanliness factor* when applied to a condenser. In this case it was the factor applied to the HEI tube bundle heat transfer coefficient in order to at least allow for the variation in Nusselt factor throughout the rows of the tube bundle (see Section 2.3.5). In the case of a heat exchanger, it has a different definition according to the appropriate HEI standard [1980]: it is the allowance applied to account for future fouling of the inside and/or outside surfaces of the tubes, separate fouling factors being selected for each surface in accordance with the TEMA standard [1988]. For instance, if the overall heat transfer coefficient of the heat exchanger is 200 Btu/(ft$^2 \cdot$ h $\cdot$ °F) and a fouling factor of 0.001°F/(Btu /(ft$^2 \cdot$ h)) is applied, then the design cleanliness factor will be 83.3%. Thus, in the design of a heat exchanger, the term *cleanliness factor* is an appropriate description of its function.

This means, however, that a clean heat exchanger will perform slightly better than design when operating in the clean state and under the original design conditions.

13.3.2 LMTD Correction Factor F

In a condenser, the vapor flows across the tube bundle, and, with a single-pass condenser, the heat flux through the tubes is a direct function of the log mean temperature difference. Even with a two-pass condenser, the vapor still flows across the tube bundle, so that the direction of the water flow is not a significant factor in determining the rate of heat transfer. However, with heat exchangers, the vapor tends to flow over and along the tubes, so that, in a two-pass heat exchanger, a portion of the water is flowing in parallel with the condensing vapor while the remainder is flowing counter to the vapor. The temperature relationships for one pass for the heat exchanger of Figure 13.2, plotted along the length of the tubes, are shown in Figure 13.5.

Underwood [1934] first derived the expression for the correction factor *F*, which must be used to adjust the LMTD calculated in Equation (13.3), so as to reflect the parallel/counterflow condition; and a later version was developed by Nagle and coworkers [Nagle 1933, Bowman 1940]. A derivation is included in Kern [1990], but the form given in the next equation and included in Hewitt at al. [1994] is the one presently used by HEI and TEMA for a double-pass heat exchanger and may be stated as follows. R is the ratio of the hot side temperature drop divided by the cold side temperature gain, thus:

$$R = \frac{T_1 - T_2}{t_2 - t_1}$$

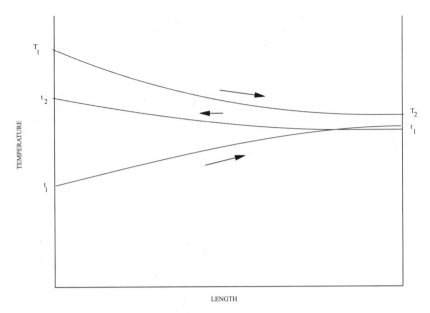

FIGURE 13.5. Temperature relationships for the two passes in the heat exchanger of Figure 13.2.

and

$$P = \frac{t_2 - t_1}{T_1 - t_1}$$

then

$$F = \frac{\sqrt{R^2 + 1} \times \ln[(1 - P)/(1 - PR)]}{(R - 1)\ln\left\{ \dfrac{2 - P[(R + 1) - \sqrt{R^2 + 1}]}{2 - P[(R + 1) + \sqrt{R^2 + 1}]} \right\}} \tag{13.5}$$

P is also known as the *temperature effectiveness*. The values of F for various values of R and P, as calculated using Equation (13.5), are plotted in Figure 13.6, which closely resembles the equivalent figures in the HEI and TEMA standards for a two-pass heat exchanger. Other derivations developed by Nagle and coworkers [Nagle 1933, Bowman 1940] show equations similar to Equation (13.5) for the values of F for heat exchangers with other configurations.

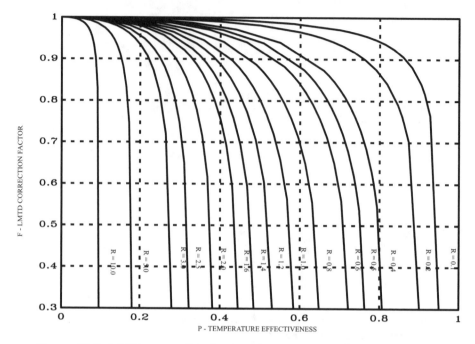

FIGURE 13.6. LMTD correction factors for a two-pass heat exchanger, for various values of *R* and temperature effectiveness *P*.

13.3.3 Calculating the Effective Heat Transfer Coefficient

In the case of the high-pressure heaters 1 and 2, rearranging Equation (13.4) and removing the cleanliness factor term (which eventually has to be calculated) allows the effective U coefficient (U_{eff}) to be obtained. If

$$Q = W_{\text{fw}}(H_{\text{out}} - H_{\text{in}})$$

$$\text{then } U_{\text{eff}} = W_{\text{fw}} \times \frac{H_{\text{out}} - H_{\text{in}}}{FA \times \text{LMTD}} \tag{13.6}$$

13.4 PREDICTED HEAT TRANSFER COEFFICIENT Upred

To determine whether the heater is performing in accordance with its design requires that the expected values of the heat transfer coefficient (U_{pred}) be calculated for the same conditions. This is the benchmark against which U_{eff} has to be compared and may be calculated from the values of F and a modified U_{ASME} value corresponding to present operating conditions.

13.4.1 Single-Tube Heat Transfer Coefficient U_{ASME}

The thermal resistance method described in Section 1.1 for calculating condenser tube heat transfer coefficient can also be used for feedwater heaters.

Shell-Side Film Coefficient

While the Nusselt equation (1.1) is appropriate for use with horizontal feed-heaters to calculate the shell-side film conductance, for feedheaters mounted vertically, as shown in Figure 13.4, Nusselt [1916] showed that the original coefficient of 0.725 in Equation (1.1) must be replaced by a value of 0.943.

Tube Wall Thermal Resistance

For the thermal resistance of the tube wall, the Kern equation (1.12) can be used. However, the TEMA tables [1988] for metal thermal conductivity show that it varies with the mean temperature. To calculate this temperature, the TEMA standard includes the following equation. Let

$$H_{fo} = \text{Nusselt film coefficient of shell-side fluid}$$
$$h_{fi} = \text{Nusselt film coefficient of tube-side fluid}$$
$$R_{fo} = \text{fouling resistance on outsides of tube (if any)}$$
$$R_{fi} = \text{fouling resistance on insides of tube (if any)}$$
$$R_w = \text{resistance of tube wall}$$
$$E_f = \text{fin efficiency (where applicable)}$$
$$T_m = \text{shell fluid average temperature}$$
$$t_m = \text{tube-side fluid average temperature}$$
$$A_o = \text{heat transfer area, tube outer surfaces}$$
$$A_i = \text{heat transfer area, tube inner surfaces}$$
$$t_{\text{mean}} = \text{tube mean metal temperature}$$

Then

$$t_{\text{mean}} = T_m - \left[\frac{\left(\dfrac{1}{h_o} + R_{fo} \right) \left(\dfrac{1}{E_f} \right) + \dfrac{R_w}{2}}{\left(\dfrac{1}{h_o} + R_{fo} \right) \left(\dfrac{1}{E_f} \right) + R_w + \left(R_{fi} + \dfrac{1}{h_i} \right) \left(\dfrac{A_o}{A_i} \right)} \right] (T_m - t_m) \quad (13.7)$$

The thermal conductivity corresponding to the value of t_{mean} should be the value used in the Kern equation (1.12).

Tube-Side Film Coefficient

To calculate the tube-side thermal resistance, the Rabas-Cane correlation of Equation (1.14) is appropriate. Alternatively, the Dittus-Boelter equation [ASME 1983] may be used, which is as follows:

$$\frac{1}{R_t} = 0.024 \left(\frac{k}{d_i} \right) \left(\frac{\rho d_i V}{\mu} \right)^{0.8} \left(\frac{C_p \mu_b}{k} \right)^{0.4} \left(\frac{\mu_b}{\eta} \right)^{0.14} \tag{13.8}$$

The nomenclature of Equation (13.8) is the same as that for the Rabas-Cane correlation, Equation (1.14), and μ_b is the viscosity of water at the bulk temperature T_b.

Overall Single-Tube Heat Transfer Coefficient

The single-tube heat transfer coefficient corresponding to current conditions can then be obtained from:

$$U_{\text{pred}} = \frac{1.0}{1/h_f + R_w + R_t} \tag{13.9}$$

13.5 HEAT EXCHANGER CLEANLINESS FACTOR

Section 13.3 showed a method for calculating the effective heat transfer coefficient U_{eff} for a heat exchanger, based upon present operating conditions and using the log mean temperature difference, surface area, and heat flux only. To compare the value of Ueff with an expected value, the value of U_{pred} obtained from Equation (13.9) must be multiplied by the LMTD correction factor F to obtain the cleanliness factor CF:

$$CF = \frac{U_{\text{eff}}}{U_{\text{pred}} F} \tag{13.10}$$

13.6 VERIFICATION WITH DESIGN AND WITH RESPECT TO LOAD

The heat exchanger relationships calculated as outlined above should be verified against actual data obtained when operating under conditions close to design. Meanwhile, Section 2.4 contained a discussion on the manner in which condenser design cleanliness factor has been found to vary with generator load; a similar relationship may also apply to heat exchangers. For these reasons, heat exchanger data should be acquired under a variety of load conditions with a clean heat exchanger, so that the actual and predicted results can be reconciled for the clean condition.

13.7 HEAT EXCHANGER CONDENSATE INVENTORY

Neundorfer [Neundorfer 1995, Lantz 1991] has shown that poor inventory control of feedwater heater liquid has been responsible for premature failure or poor heater performance. If the water level in a feedwater heater is too low, then steam enters the drain cooler section, causing tube or drain baffle erosion, and can have a negative impact on heat rate. Conversely, if the water level is too high, the surface area of the condensing zone is reduced, which again degrades the unit's heat rate and can also cause tube failures in the desuperheating zone. Thus, proper control consists of ensuring that the water level is held within close tolerances, not only during steady state but also during load transients. A method for achieving this result was outlined by Lantz [1991].

13.8 AUXILIARY HEAT EXCHANGERS IN NUCLEAR PLANTS

In a nuclear power plant there are several auxiliary heat exchangers which are supplied with cooling water from the service-water system, and many are associated with plant safety systems. During an emergency, it is important that these heat exchangers be in proper condition to remove the amount of heat for which they were designed. Unfortunately, excessive fouling was discovered during the 1980s in many such heat exchangers at several different plants. This led the Nuclear Regulatory Commission [1989] to issue Generic Letter 89-13, which required that all holders of nuclear operating licenses regularly evaluate the condition of the emergency service-water systems to ensure that they can remove the heat loads specified in the postulated event. Among the heat exchangers at risk are those associated with:

- Generator coolers
- Diesel engine jacket coolers
- Component coolers
- Control room heating and ventilation systems
- Air compressor inlet and interstage coolers
- Motor control center cooling cells

Unfortunately, since most of these heat exchangers are associated with emergency systems, they are normally shut off, or else the cooling-water flow through them is small. Thus, because they are, it is hoped, normally in this state, these heat exchangers have a great tendency to foul and/or corrode and, for the same reasons, are difficult to monitor in situ.

Attempts were therefore made to simulate the postulated events by other means. The late Richard C. Schwarz designed a set of portable condensers [Putman et al. 1997] which have been used to simulate condensing conditions and allow the heat transfer of the tubes in the vessel to be continuously monitored; some of this apparatus was described in Chapter 10. When supplied with the same service water as the real heat exchanger and operated under the same conditions, the tubes can be removed periodically and examined. Such models allow judgment of the condition of the tubes in the actual heat exchanger installed to safeguard the unit, but without interfering with its operation or safety. However, such simulations will not reproduce the conditions on water-to-water heat exchangers, although models of these have been constructed to measure heat transfer, the water in one path being heated externally.

In practice, at least two characteristics have to be monitored on these safety-related heat exchangers:

1. Whether the heat transfer coefficient of the tubes is less than its design value, including any fouling resistances assigned to the inner and outer tube surfaces
2. If the tubes are fouled, whether the obstruction to flow is such that it will hold the flow rate below the design value

If some flow can be allowed to pass through the unit for test purposes and an appropriate set of instrumentation has been provided, then it should be possible to compute the heat transfer coefficient of the tubes, due allowance being made for any calibration coefficients and any observed tendency for the design or reference heat transfer coefficient to vary with water flow rate. In the case of flow, comparison of the pressure drop under lower flow conditions against the design pressure drop under full flow conditions should permit a judgment to be made regarding the pump discharge pressure required under full flow. This can then be compared with the design pump discharge pressure at full flow.

Many nuclear plants were designed some years ago, so that some of their heat exchangers may have been built to earlier standards and practices. A critical analysis of many design data sheets has revealed data inconsistencies, and it is prudent to reexamine the assumptions incorporated in the original design, together with the accuracy of its implementation as embodied in the heat exchanger design data sheet or specification.

Appendix A

MATHEMATICAL PROCEDURES

Within the body of this book there are references to a number of mathematical procedures which are reviewed in this appendix: multiple linear regression analysis, and several numerical analysis techniques, chief among them the Newton-Raphson method.

A.1 MULTIPLE LINEAR REGRESSION ANALYSIS

Multiple linear regression analysis was originally conceived as a means for statistically analyzing experimental data. Draper and Smith [1968] provide a thorough outline of the fundamental theory, and reference should be made to their work. Consider a situation in which a result R can be measured corresponding to the measured values of a set of n independent variables $X_1, X_2, \ldots, X_n$. When m complete sets of (X, R) data are presented to a multiple linear regression analysis program, it will calculate the values of coefficients $a_0, a_1, \ldots, a_n$ in the following expression:

$$R = a_0(1.0) + a_1 X_1 + a_2 X_2 + \cdots + a_n X_n \qquad (A.1)$$

these values of a providing the best (minimum least-squares) fit of this expression to all the data sets.

In matrix notation, if the values of R in the m data sets are arranged as a vector $\mathbf{Y}$ while the values of X are arranged as an (X_{n+1}, m) matrix $\mathbf{Z}$ in which the first column contains the value $X_0 = 1.0$ for all data sets, and if vector $\mathbf{A}$ is to contain the elements $a_0, a_1, \ldots, a_n$, then:

$$\mathbf{A} = (\mathbf{Z}'\mathbf{Z})^{-1}\mathbf{Z}'\mathbf{Y} \qquad (A.2)$$

In matrix $\mathbf{Z}$, the number of rows should be at least equal to the number of variables plus 1, or $m \geq (n+1)$; $\mathbf{Y}$ will contain m elements; and the vector $\mathbf{A}$ will contain $n + 1$ elements. Draper and Smith also indicate several criteria for judging the quality of the fit of the calculated array $\mathbf{A}$, but these are not of concern when regression analysis is being used within a system for monitoring condenser performance. The principal software modules required to resolve the value of vector $\mathbf{A}$ include a module to transpose a matrix, a module to multiply two matrices together, and a module to calculate the inverse of a matrix.

The main use of regression analysis in condenser performance modeling is to generate the coefficients in polynomial relationships which take the form:

$$R = a_0(1.0) + a_1 X + a_2 X^2 + a_3 X^3 \tag{A.3}$$

Here R is again the result, while X is the value of a *single* independent variable associated with each value of R. Generating the **Z** matrix in Equation (A.2) so as to reflect the polynomial form of Equation (A.3) requires making the substitutions $X_1 = X$, $X_2 = X^2$, and $X_3 = X^3$. Equation (A.2) can now be used to solve for vector **A**, which will contain a new set of values $a_0, a_1, \ldots, a_n$ that best satisfy the polynomial of Equation (A.3).

A.2 NUMERICAL ANALYSIS

With equations of the form $Y = f(X)$, the value of Y can be calculated directly from the selected value of X. However, if it is desired to calculate the value of X corresponding to the value of Y having the value of interest (Y^*), it will be necessary to use some type of numerical analysis method in order to resolve the value of X when $Y = Y^*$, unless the equation can be directly converted to the form

$$X = f(Y)$$

The literature contains several simple methods for solving $Y^* = f(X)$, described in the following subsections.

A.2.1 Fibonacci Search

A Fibonacci search can be useful if it is known with certainty that the solution value of X^* lies within a range L of Y whose upper and lower boundaries (Y_{max}, Y_{min}), together with the associated values of X_{max} and X_{min}, have been determined. Wilde [1964] shows the economy of a *search by golden section*, in which a first estimate of the solution is obtained by dividing the range L by the golden mean $\tau = 1.618033989$ and evaluating the value of X corresponding to the value of:

$$\text{Either} \left(X^* = X_{max} - \frac{L}{\tau} \right) \quad \text{or} \quad \left(X^* = X_{min} + \frac{L}{\tau} \right) \tag{A.4}$$

The error corresponding to the value of X^* chosen for this first estimate is $(Y - Y^*)$, and so, depending on its sign, the upper and lower limits of either the longer or shorter section of L become the new values of X_{max} and X_{min}. A new value of X^* is calculated according to Equation (A.4) using the reduced length of L, and the new error $(Y - Y^*)$ is then calculated; the search process continues until the last calculated error value lies within the desired tolerance.

A.2.2 Regula Falsi Search

Kreyszig [1973] outlines the regula falsi method, which, again, requires a knowledge of the range of X within which the solution value $(Y = Y^*)$ will be found. By calculating the values Y_{max} and Y_{min} corresponding to the values of X_{max} and X_{min}, a first approximation of X^* can be calculated by using triangulation:

$$X^* = \frac{X_{min}Y_{max} - X_{max}Y_{min}}{Y_{max} - Y_{min}} \tag{A.5}$$

A new value of Y^* can be calculated from X^* together with the error $(Y - Y^*)$, and so, depending on the sign of the error, the newly calculated values of X^* and Y^* will be substituted for either the present upper or the present lower limit values of X and Y—i.e., for (X_{max}, Y_{max}) or for (X_{min}, Y_{min}). The updated range limits of X and Y will now be substituted in Equation (A.5) and new values of X^* and Y^* calculated, together with a new value of the error $(Y - Y^*)$. The process continues until the error falls within the desired tolerance.

A.2.3 Newton-Raphson Numerical Search Technique

Kreyszig [1973, p. 764] cites a numerical analysis technique developed by Joseph Raphson [1697] that has been used in a variety of ways within programs designed to monitor condenser performance. Figure A.1 demonstrates the principles involved. In this example, the value of X is sought for the equation $Y = X^3 - 2X - 5 = 0$. The desired tolerance is an error less than 10^{-8}. X is chosen for first estimate to have a value of 1.5 (point A), and it is seen that the error at point A in Figure A.1 is -4.625 while the slope is 4.75. To obtain an improved estimate of the solution, the value of X should be changed by an amount equal to the quotient: $(-error/slope)$. In this case, the change in X (DELX1, in the figure) would be 0.973, so that the value of X which will provide an improved estimate of Y is now 2.473 (point B), the error at this point being 5.189 and the slope 16.357.

The change in X (DELX2) which will improve the estimate is now $-5.189/16.357 = -0.317$, the updated value of X now being 2.1564 (point C); the error at this point is 0.715 and the slope 11.95. The change in X (DELX3) which will improve the estimate further is now $-0.715/11.95 = -0.06$, the updated value of X now being 2.097 (point D). Using the slope and error calculated for point D, the next iteration will move the value of X to 2.0945, which is the solution value for $Y = 0$, because the error is now less than the desired threshold of 10^{-8}.

The program displayed in Table A.1 summarizes the Newton-Raphson numerical search method. First, select a value of X as a first estimate, and then calculate the error and slope of the curve at that point. By changing the previous value of X by an amount equal to $-error/slope$, or setting $X = X - (error/slope)$, the error calculated at the next iteration will be reduced. Recalculate the values of error and slope at this new point and update the value of X. Continue until the value of the error falls below some maximum value.

Note that $Y = f(X)$ is assumed to be a monotonic equation and that a unique

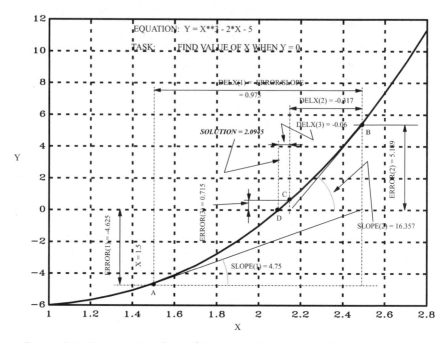

FIGURE A.1. Demonstration of Newton-Raphson method.

solution to the problem is also assumed. However, this is not always true. For instance, a third-order polynomial equation will have three roots, and if a search technique is to be used to establish the values of these roots, the associated ambiguity must be recognized. In such cases, a plot of the equation should be developed by calculating Y for a wide range of X values and analyzing the plot to define what should be the rules for a successful search. Even when monotonicity is present, the search procedure can often be speeded up by first calculating X over a wide range of Y, noting when the error $(Y - Y^*)$ changes sign, and so establishing a reduced range of X within which to conduct the final search.

A.3 THE NEWTON-RAPHSON TECHNIQUE USED IN CONDENSER PERFORMANCE MODELS

A condenser performance model consists of a set of nonlinear simultaneous equations which, being nonlinear, can best be solved using the Newton-Raphson procedure. If there are n variables, there must also be n equations, some of which define the boundary conditions on which the model must converge. For instance, to model the behavior of a clean single-compartment condenser in a fossil fuel plant requires a matrix having 5 equations and 5 variables, the boundary conditions consisting of generated power, cooling-water inlet temperature, and flow

Table A.1. Program to Demonstrate the Newton-Raphson Principle

```
C          PROGRAM TO DEMONSTRATE THE NEWTON-RAPHSON PRINCIPLE
C
      INTEGER*2   I
      REAL*8   X,ERRX,DFDX,DELX
C
C   FUNDAMENTAL EQUATION:   Y = X**3 - 2.0 * X - 5.0
C
C   TASK:    FIND THE VALUE OF X WHEN Y = 0
C
C   ASSIGN INITIAL VALUE TO 'X' AND CALCULATE ASSOCIATED ERROR
C
      X = 1.5
        ERRX = X**3 - 2.0*X - 5.0
          WRITE(*,101)
           WRITE(*,100) X,ERRX
              WRITE(*,102)
                WRITE(*,103)
                I = 0
C
C   LOOP TO IMPLEMENT NEWTON-RAPHSON PRINCIPLE
C
           DO WHILE(ABS(ERRX) .GE. 1.0E-08)
              I = I + 1
                 DFDX = 3.0 * X**2 - 2.0
                   DELX = - 1.0 * ERRX / DFDX
                     X = X + DELX
                        ERRX = X**3 - 2.0*X - 5.0
                           WRITE(*,104) I,X,ERRX,DFDX,DELX
           END DO
        WRITE(*,101)
C
C   SOLUTION
C
      WRITE(*,200) X
C
C   PRINTOUT FORMATS
C
  100 FORMAT(1H ,' Initial value of X = ',F15.9,'  Initial Error = ',
     + F15.9 // )
  101 FORMAT(20X / )
  102 FORMAT(1H ,' Iteration                            Partial')
  103 FORMAT(1H ,'    No.          X              Error      Differential
     +      Delta X' / )
  104 FORMAT(1H ,I5,4F15.9)
  200 FORMAT(1H ,'Solution to equation : X**3 - 2.0*X - 5.0 = 0.0   :
     +      X = ',F12.9  //)
      END
```

(see Table A.2). At model convergence, all equations take the form

$$f(X_1) + f(X_2) + \cdots + f(X_n) = 0 \qquad\qquad (A.6)$$

Table A.2. Newton-Raphson Data Structure for Single-Compartment Condenser

Change vector **G**					
ΔTFLOW	ΔTIN	ΔEXH1	ΔSTMP1	ΔTOUT1	
1					-err(1)
	1				-err(2)
		1	$\partial f(3)/\partial STMP1$		-err(3)
$\partial f(4)/\partial TFLOW$	$\partial f(4)/\partial TIN$	$\partial f(4)/\partial EXH1$		$\partial f(4)/\partial TOUT1$	-err(4)
$\partial f(5)/\partial TFLOW$	$\partial f(5)/\partial TIN$		$\partial f(5)/\partial STMP1$	$\partial f(5)/\partial TOUT1$	-err(5)
Matrix of partial differentials					Error vector **E**

Once the set of equations has been defined, the n variables in the equation set are assigned initial values, which might be the present set of operating conditions. Using this initial set of variable values, the sum of $f(X_i)$, $i = 1, \ldots, n$, can be evaluated for each equation, and, since it is unlikely that convergence has already occurred, each of these sums will have a nonzero value. These sums can be stored in a vector **E** as the *negatives* of their values.

An $n \times n$ matrix **F** can now be constructed from each equation, the matrix elements consisting of the partial differentials of each variable in each equation. These partial differentials can be determined in two ways:

1. By inspection of the equation. For example, if $f(X_{i,j}) = a_1 X_{i,1}$, then the partial differential of element $X_{i,j}$ will be a_1, which will be stored in element $F_{i,j}$.
2. Alternatively, the partial differential of $X_{i,j}$ with respect to equation j can be calculated by perturbing $X_{i,j}$ by a small amount, calculating the effect of that perturbation on the sum for equation j, and then dividing the change in the sum by the perturbation.

Assume a vector **G** which is to contain the *changes* to be made in the value of each variable in order for the solution to approach closer to convergence, convergence being defined as having occurred when all values contained in **G** fall below the threshold value assigned to each variable. Considering the tableau contained in Table A.2, the basic relationship between the matrix and the vectors is **GF = E**, from which the values for array **G** can be calculated using:

$$\mathbf{G} = \mathbf{EF}^{-1} \tag{A.7}$$

After updating the set of variables by applying to each the corresponding 'change' value contained in the elements of the $\mathbf{G}$ vector, new error values are calculated for each equation and stored in array $\mathbf{E}$, while all the partial differentials are also recalculated, using the updated values of each variable. The procedure defined in Equation (A.7) is performed and a new set of change values calculated. If the absolute values of all changes fall below the assigned thresholds, then convergence is declared and the current set of X values becomes the solution. Otherwise, if the change to be made in any variable should fall outside its assigned tolerance, the iterative procedure will continue until convergence takes place. In practice, convergence for a condenser matrix will occur in fewer than 10 iterations. Should a larger number of iterations be required, it is possible that some tolerances will need to be adjusted or the set of equations critically examined.

There are strong similarities between this method of solving a set of nonlinear simultaneous equations and the Newton-Raphson search technique outlined in Section A.2.3. In both cases, in order to cause the model to move closer to convergence, a better variable, or set of variables, can be obtained by changing the present values of X_i, $i = 1, \ldots, n$, by amounts equivalent to the negative of the error divided by the slope or, in the case of condenser models, the matrix equivalent of slope considered as the set of partial differentials.

Appendix B

PROPERTIES OF SALINE WATER

The following are FORTRAN subroutines for calculating the density and specific heat of saline water, based on Figures A-2 and A-4 of PTC 12.2-1983 [ASME 1983].

FUNCTION FOR DENSITY OF SALINE WATER

```
      REAL*4    FUNCTION dens_sea_mix(T,SEAPERC)
C
C     T      = COOLING WATER TEMPERATURE - Deg.F
C     SEAPERC = PERCENTAGE OF SEA WATER IN COOLING WATER
C               100.0  = SEA WATER ONLY
C                 0.0  = FRESH WATER ONLY
C
      REAL*4    T,SEAPERC,CDENS(4),CSEADEN(4)
      REAL*4    FRESHDEN,SEADEN,RANGE
C
      DATA CDENS  /62.087,2.2519E-02,-3.3873E-04,1.0579E-06/
      DATA CSEADEN/63.77,2.0328E-02,-3.5472E-04,1.3248E-06/
C
      FRESHDEN = CDENS(1) + CDENS(2)*T + CDENS(3)*T**2 + CDENS(4)*T**3
         SEADEN = CSEADEN(1) + CSEADEN(2)*T + CSEADEN(3)*T**2
              +           + CSEADEN(4)*T**3
        RANGE  = SEADEN - FRESHDEN
           dens_sea_mix = FRESHDEN + (SEAPERC/100.) * RANGE
      RETURN
      END
```

FUNCTION FOR SPECIFIC HEAT OF SALINE WATER

```
        REAL*4    FUNCTION spht_sea_mix(T,SEAPERC)
C
C       T       = COOLING WATER TEMPERATURE - Deg.F
C       SEAPERC = PERCENTAGE OF SEA WATER IN COOLING WATER
C               100.0  = SEA WATER ONLY
C                 0.0  = FRESH WATER ONLY
C
        REAL*4    T,SEAPERC,CSPHT(4),CSEASPHT(2)
        REAL*4    FRESHSPH,SEASPHT,RANGE
C
        DATA CSPHT    /1.0244,-7.1715E-04,6.1796E-06,-1.6397E-08/
        DATA CSEASPHT/9.5062E-01,6.2292E-05/
C
        FRESHSPH = CSPHT(1) + CSPHT(2)*T + CSPHT(3)*T**2 + CSPHT(4)*T**3
         SEASPHT = CSEASPHT(1) + CSEASPHT(2)*T
           RANGE = FRESHSPH - SEASPHT
              spht_sea_mix = SEASPHT + (1.0 - SEAPERC/100.) * RANGE
        RETURN
        END
```

REFERENCES

Anderson, S., et al. (1990). "Mechanical Manganese Dioxide Removal in Stainless Steel Condenser Tubes," EPRI Service Water Systems Cleaning Technology Seminar, Palo Alto, CA, February 21–22.

Andreone, C. F., and S. Yokell (1998). *Tubular Heat Exchanger Inspection, Maintenance and Repair*, McGraw-Hill, New York.

Antoine, J., and G. Smith (1991). "Project DISCOVERY—Entergy Operations' Program for Evaluating Water Treatment Proposals," International Water Conference, Pittsburgh, PA, October.

ASM (1987). *Metals Handbook: Corrosion*, 9th ed., ASM, Metals Park, OH.

ASME (1983). "Code on Steam Condensing Apparatus," ASME Power Test Code PTC.12.2-1983, American Society of Mechanical Engineers, New York.

ASME (1985). "Measurement Uncertainty," Performance Test Code PTC.19.1-1985, American Society of Mechanical Engineers, New York.

ASME (1988). "Procedures for Routine Performance Tests of Steam Turbines," ASME Power Test Code PTC 6S Report-1988, American Society of Mechanical Engineers, New York.

ASME (1993). Steam Tables, 6th ed., American Society of Mechanical Engineers, New York.

ASME (1998). "Performance Test Code for Steam Condensers," ASME PTC.12.2-1998, American Society of Mechanical Engineers, New York.

Bell, K. (1994). "Water Guns Used to Clean 55,000 Condenser Tubes," *Inside Nuclear*, November, pp. 5, 11, publ.PECO, Philadelphia, PA.

Bowman, R. A., A. C. Mueller, and W. M. Nagle (1940). " Mean Temperature Difference in Design", *ASME Transactions*, Vol. 62, pp. 283–294.

Carpenter, B. H., and H. C. Sweeney (1965). "Process Improvement with ' Simplex' Self-Directing Evolutionary Operation," *Chemical Engineering*, July 5, pp. 117–126.

Colborn, C.E. (1923). "Determining the Economical Interval between Cleanings of Condenser Tubes," *Power*, November, pp. 803–805.

Compton, K. G., and E. F. Corcoran (1974). "The Discharge of Copper Corrosion Products from Steam Electric and Desalination Plants and Its Effect on the Nearby Ecosystem," INCRA Project No. 215, June 1.

Compton, K. G., and E. F. Corcoran (1976). "Concentrations, Form and Deposition of Copper Corrosion Products in Recirculating Cooling Systems for Power Plants," INCRA Project No. 253, June 1.

Conco (1985). *Bulletin*, Vol. 1, No. 4, Fall, Conco Systems, Inc.

Conley, E. F. (1983). "Corrosion Monitoring of Condensing Systems," EPRI Symposium on State-of-the-Art Condenser Technology, June, Orlando, FL.

Davenport, S. D., W. J. Davis, and K. A. Selby (1992). "A Service Water System Monitoring Skid Designed to Test the Impact of Chemical Addition," *Proceedings, EPRI Service Water System Reliability Improvement Seminar*, June, Daytona Beach, FL.

DeMoss, T. B. (1995). "Finding Lost Megawatts," *Power Engineering*, Vol. 99, No.12 (December), pp. 21–24.

Draper, N. R., and H. Smith (1968). *Applied Regression Analysis*, John Wiley & Sons, New York.

Durum, W.H. and Joseph Haffty (1963). "Implications of the Minor Element Content of Some Major Streams of the World," *Geochemica*, Vol. 27, pp. 1–11.

EPA (1988). *Water Quality Standards Criteria Summaries: A Compilation of State/Federal Criteria*, U.S. Environmental Protection Agency, Office of Water Regulations and Standards, Washington, DC 20460, September.

EPA (1994). "Methods for the Determination of Metals in Environmental Samples," EPA-600/R-94/111, May, U.S. Environmental Protection Agency, Environmental Monitoring Systems Laboratory, Cincinnati, OH.

EPA (1995). "Method 1640: Determination of Trace-Elements in Ambient Waters by On-Line Chelation Preconcentration and Inductively Coupled Plasma-Mass Spectrometry," EPA 821-R-95-033, U.S. Environmental Protection Agency, Office of Water Regulations and Standards, Washington, DC 20460, April.

EPRI (1982), "Steam Plant Surface Condenser Leakage Study Update", EPRI Report NP-2062, May.

EPRI (1988). "Condenser Leak Detection Guidelines Using Sulfur Hexafluoride as a Tracer Gas," Report CS-6014, Project 1689-20, September.

EPRI (1989). "Condenser Leak Detection by Using SF_6 as a Tracer Gas," Technical Brief TB.GSZ.89.11.89.

EPRI (1997). "In-Situ Coating of Condenser Tubes as an Alternative to Retubing," EPRI Report TR-107068, September 1997.

Florida DEP (1992). Standard Operating Procedures, Florida Department of Environmental Protection DEP-QA-001/92.

Frisina, V. C., L. N. Carlucci, et al. (1989). "Advanced Predictive Performance Modelling for Condensers and Feedwater Heaters," *Proceedings, EPRI 1989 Heat Rate Improvement Conference*, Knoxville, TN, September, pp. 7B-37 to 7B-51.

Fromberg, R. R., and D. A. Leach (1997). "On-Line Condenser Performance Monitor at PG&E' s Pittsburgh Power Plant Unit 7," *Proceedings, 1997 Joint Power Generation Conference*, Denver, ASME, New York, PWR-Vol. 32, Vol. 2, pp. 467–474.

Garey, J. F., J. L. Tsou, and D. Wiebe (1996). "On-Line Monitoring of Performance Losses Due to Fouling," *Proceedings, EPRI Condenser Technology Conference*, Boston, August, pp. 11-1 to 11-17.

Griffin, C. M. (1931). Tube Cleaner, U.S. Patent No. 1,814,752, July 14.

Griffin, C. M. (1939). Tube Cleaners, U.S. Patent No. 2,170,997, August 29.

Griffin, C. M. (1947). Fluid propelled articulated scraper for cleaning tubes, U.S. Patent No. 2,418,509, April 8, 1947.

Griffin, C. M. (1956). Tube Cleaner, U.S. Patent No. 2,734,208, February 14.

HEI (1980). *Standard for Power Plant Heat Exchangers*, 1st ed., Heat Exchange Institute, Cleveland.

HEI (1984). *Standards for Steam Surface Condensers*, 8th ed., Heat Exchange Institute, Cleveland.

HEI (1989). Addendum to *Standards for Steam Surface Condensers*, 8th ed., Heat Exchange Institute, Cleveland.

HEI (1995). *Standards for Steam Surface Condensers*, 9th ed., Heat Exchange Institute, Cleveland.

Henderson, C. L., and J. M. Marchello (1969). "Film Condensation in the Presence of a Non-Condensable Gas," *Journal of Heat Transfer*, Vol. 91, August, pp. 447–450.

Hewitt, G. F., G. L. Shires, and T. R. Bott (1994). *Process Heat Transfer*, CRC Press, Boca Raton, FL.

Holt, A. E., G. M. Light, et al. (1993). "Conversion of NDE Inspection Signals into Aural Signals for Improving Reliability," *Proceedings ICOPE-93*, Tokyo, September, pp. 225–230.

Hovland, A. W. (1978) " Effective Condenser Cleaning Improves System Heat Rate", *Power Engineering*, January 1978, pp.49-50.

Hovland, A. W. (1990). "The Economics of Mechanical Condenser Cleaning," presented at IJPGC, Boston, October.

Hovland, A. W., D. A. Rankin, and E. G. Saxon (1988). "Heat Exchanger Tube Wear by Mechanical Cleaners," *Proceedings, Joint Power Generation Conference*, Paper 88-JPGC/Pwr-8, Philadelphia, September 25–29, pp. 1-4.

IPCC (1995). *Climate Change 1994: Radiative Forcing of Climate Change and an Evaluation of the IPCC IS92 Emission Scenarios*, Cambridge University Press, Cambridge, UK.

Katragadda, G., J. T. Si, and G. P. Singh (1997). "A Fuzzy Inference Adviser for the Investigation of Condenser Performance Deficiencies," *Proceedings, IJPGC*, Denver, ASME, New York, PWR-Vol. 32, Vol. 2, pp. 307–312.

Kaul, T., and E. Willey (1990). "Implementation of Computerized Data Management for Balance of Plant ECT Inspections," EPRI NDE BOP Workshop, New Orleans, August.

Keller (1992). *"Confined Spaces—A Training Program for Employees*, J. J. Keller & Associates, Neenah, WI.

KEPCO (1992). Internal Report prepared by KEPCO Technical Research Institute, on Condenser #2 at Yong Nam Thermal Power Plant, September.

Kern, D. Q. (1958). "Mathematical Development of Loading in Horizontal Condensers," *AIChEJ*, Vol. 4, p. 157.

Kern, D. Q. (1956). *Process Heat Transfer*, 2nd ed., McGraw-Hill, New York.

Kim, Dong Sup and R. E. Putman (1993). "The Management of Condenser Tube

Pitting at #2 Unit, Yong Nam Thermal Power Plant, to Improve Unit Performance and Availability," *Proceedings of IJPGC*, Kansas City, October, PWR-Vol.23, pp. 43–48.

Koch, J., C. J. Haynes, and Y. Mussalli (1996). "Economic Scheduling of Condenser Cleaning Based on Computerized Thermal Performance Monitoring," EPRI Condenser Technology Conference, Boston, pp. 3-110 to 3-119.

Kreyszig, E. (1973). *Advanced Engineering Mathematics*, John Wiley & Sons, New York.

Lantz, J. B. (1991). Method and Apparatus for Optimization of Feedwater Heater Liquid Level, U.S. Patent No. 5,012,429, April 30.

Lewis and Whitfield (1974). "The Biological Importance of Copper in the Sea," International Copper Research Association, Project 223, April.

Lewitt, E. H. (1953). *Thermodynamics Applied to Heat Engines*, 5th ed., Pitman & Sons, London, pp. 274–289.

Little, B., et al. (1998). "The Role of Biomineralization in Microbiologically Influenced Corrosion," Paper No. 294, Corrosion 98, 17 pp.

Longbottom, J. E., et al. (1994). "Determination of Trace Elements in Water by Inductively Coupled Plasma-Mass Spectrometry: Collaborative Study," *Journal of AOAC International*, Vol. 77, pp. 1004–1023.

Ma, R. S. T., and N. Epstein (1981). "Optimum Cycles for Falling Rate Processes," *The Canadian Journal of Chemical Engineering*, Vol. 59, October, pp. 631–633.

March, P. A., and C. W. Almquist (1987). "Techniques for Monitoring Flowrate and Hydraulic Fouling of Main Steam Condensers," Tennessee Valley Authority, Engineering Laboratory, Report No. WR28-1-41-106, February 1987.

Morris, L., et al. (1985). "Manganese Deposits in Utility Condensers—Experience Report," EPRI Condenser Biofouling Control Symposium, Lake Buena Vista, FL, June 18–20.

Mussalli, Y., G. E. Hecker, C. Cooper, et al. (1991). "Improved On-Line Condenser Cleaning Ball Distribution," *Proceedings, ASME Power Conference, Practical Aspects and Performance of Heat Exchanger Components and Materials*, Minneapolis, PWR-Vol. 14, pp. 51–56.

Nagle, W. M. (1933). "Mean Temperature Differences in Multipass Heat Exchangers ", *Industrial Engineering Chemistry*, Vol. 26, pp. 604–608.

NALCO (1998). Private communication from NALCO Chemical Corporation, March.

Neundorfer, M., and D. K. Frerichs (1995). Advanced Feedwater Heater Control, *Proceedings, American Power Conference*, Chicago, Vol. 57, pp. 759–763.

New England Power Service (1989). "Condenser Tube Plug Evaluation," Report, September 29.

Nuclear Regulatory Commission. "Service Water System Problems Affecting Safety Related Equipment," Generic Letter 89-013; and Supplement 1 to GL 89-013.

Nusselt, W. (1916). "Die Oberflaschenkondensation des Wasserdampfes," *Zeitschrift des V.D.I.*, Vol. 60, pp. 541–546, 569–575.

OSHA (1999), Code of Federal regulations, 29 CFR 1910.146, Occupational Safety and Health Standards, Permit-required confined spaces, Revised as of July 1, 1999, pp.448-470.

Putman, R. E. (1994). "The On-Line Calculation of Condenser Performance and Optimum Cleaning Interval," 10th International Heat Transfer Conference, Brighton, UK, August, pp. 207–212.

Putman, R. E. "An Improved Method of Estimating the Performance of Condensers with Both Single and Multiple Compartments," EPRI Nuclear Plant Performance Improvement Seminar, Albuquerque, August 23–24.

Putman, R. E. (1997a). "Calculating an Optimum Condenser Cleaning Schedule with Either Linear or Non-Linear Fouling Models," IJPGC 1997, Denver, November 3–5.

Putman, R. E. "Performance Monitoring of Three-Compartment Condensers in Nuclear Plants," EPRI Nuclear Plant Performance Improvement Seminar, San Antonio, TX, August 23–24.

Putman, R. E., and Joseph W. Harpster (2000), "The Economic Effects of Condenser Back Pressure on Heat Rate, Condensate Subcooling and Feedwater Dissolved Oxygen", *Proceedings, IJPGC 2000*, Miami.

Putman, R. E., and M. J. Hornick. "Using Turbine Thermal Kit Data to Benchmark Condenser Performance Calculations," *Proceedings, IJPGC 1998*, Baltimore, pp. 511–524.

Putman, R. E., and D. C. Karg. "Monitoring Condenser Cleanliness Factor in Cycling Plants," *Proceedings, IJPGC 1999*, PWR-Vol. 34, San Francisco, ASME, New York, pp. 133–142.

Putman, R. E., and G. E. Saxon, Jr. "A Newton-Raphson Method for Calculating Condenser Performance Based on ASME Single Tube Heat Transfer Data," *Proceedings, EPRI Heat Rate Improvement Conference*, Dallas, TX, May 22–24, pp. 17-1 to 17-24.

Putman, R. E., J. Kocher, D. R. Papanek, and M. J. Rohall (1994). "The Better Use of Eddy Current Data to Improve Heat Exchanger Maintenance Management," 3rd EPRI BOP Heat Exchanger Workshop, Myrtle Beach, NC, June 6–8.

Putman R. E., G. E. Saxon, Jr., and M. Rohall. "Cooling Water Monitors for the Study of Corrosion, Fouling and Heat Transfer in Condensers and Heat Exchangers," EPRI Service Water Systems Reliability Improvement Seminar, Denver, June 4–5, Session III.

Qi Zhao et al. "Dropwise Condensation of Steam on Vertical and Horizontal U-Type Tube Condensers," *Proceedings, International Heat Transfer Conference*, Brighton, UK, pp. 117–121.

Rabas, T. J., and D. Cane (1983). "An Update of Intube Forced Convection Heat Transfer Coefficients of Water," *Desalination*, Vol. 44, pp. 109–119, Elsevier, Holland.

Rabas, T. J., C. B. Panchal, et al. (1991). "Comparison of Power-Plant Condenser Cooling-Water Fouling Rates for Spirally-Indented and Plain Tubes," ASME Heat Transfer Conference, HTD-Vol. 164, Fouling and Enhancement Interactions, Minneapolis.

Raiffa, H. (1970). *Decision Analysis: Introductory Lectures on Choices under Uncertainty*, Addison-Wesley, Reading, MA.

Raphson, J. (1697). *Analysis Aequationum Universalis*, 2nd ed., Royal Philosophical Society, London.

Rhodes, N., and R. J. Bell (1998). "Condenser Configuration and Performance Analysis Using Computational Fluid Dynamics," *Proceedings, IJPGC 1998*, Baltimore, pp. 459–471.

Rifert, V. G., A. I. Sardak, et al. (1989). "Heat Exchange at Dropwise Condensation in Heat Exchangers of Desalination Plants," *Desalination*, Vol. 74, pp. 373–382.

Robinson, E. L. (1933). "Leaving-Velocity and Exhaust Loss in Steam Turbines," *ASME Transactions*, Paper FSP-56-10, pp. 515–526.

Roosenberg, D. J. "Greening and Copper in the American Oyster in the Vicinity of a Steam Electric Plant," *Chesapeake Science*, Vol. 10, No. 3-4, p. 241.

Saxon, E. G., M. A. Janke, and A. W. Hovland (1989). *Condenser Tube Plug Evaluation Utilizing an Off-Line Tube Plug Tester*, ASME Book No. H00514-1989, American Society of Mechanical Engineers, New York, pp. 29–32.

Saxon, G., Jr., and R. E. Putman (1995). "Improved Condenser Performance Can Recover Up to 25 MW Capacity in a Nuclear Plant," *Proceedings, EPRI Nuclear Plant Performance Improvement Seminar*, Albuquerque, August 23–24, pp. 265–278.

Saxon, G., Jr., and R. E. Putman (1996). "Condenser Performance Recovery in Nuclear Power Plants," *Proceedings, IJPGC 1996*, Houston, ASME, Vol. NE-20, Vol. 3, pp. 41–46.

Saxon, G. E. (1981). Tube Cleaner, U.S. Patent No. 4,281,432, August 4.

Saxon, G. E. (1990). "Copper Alloy Tube Cleaning," TPT-, Vol.1, No.1, September-October 1990, publ. INTRAS Inc., Danbury, CN, pp. 28–30.

Saxon, G. E., R. E. Putman, and R. Schwarz (1996). "Diagnostic Technique for the Assessment of Tube Fouling Characteristics and Innovation of Cleaning Technology," *Proceedings, EPRI Condenser Technology Conference*, Boston, August, pp. 14-1 to 14-14.

Saxon, G. J. (1992). Tube Cleaning Tool for Removal of Hard Deposits, U.S. Patent No. 5,153,963, October 13.

Saxon, G. J., and J. Krysicki (1998). Easy Insert Tube Cleaner, U.S. Patent No. 5,784,745, July 28.

Schwarz, R. C. (1988). "The Operation of Auxiliary Heat Exchangers for Optimum Useful Life and Thermal Efficiency," Joint ASME/IEEE Power Generation Conference, Philadelphia, September 25–29, Paper 88-JPGC/Pwr-48.

Schwarz, R., R. Tombaugh, and D. Papanek (1992). "The Evaluation of Water Treatment Service Vendor Proposals through the Use of a Model Cooling System," *Proceedings, 1992 EPRI Service Water System Corrosion Seminar*, Clearwater Beach, FL, April 21–24.

Sedricks, J. A. (1979). *Corrosion of Stainless Steels*, John Wiley & Sons, New York.

Short, B. E., and H. E. Brown (1951). "Condensation of Vapors on Vertical Banks of Horizontal Tubes," *Proceedings of the General Discussion on Heat Transfer*,

Institution of Mechanical Engineers and ASME, Section I, 27, London, pp. 27–31.

Silver, R. S. (1963–1964). "An Approach to a General Theory of Surface Condensers," *Proceedings, Institution of Mechanical Engineers*, Vol. 178, Part 1, No. 14, pp. 339–376.

Silvestri, G. J., Jr. (1995). communication dated May 4.

Silvestri, G. J., Jr. (1997). Private communication.

Simmonds, P. G., A. J. Lovelock, and J. E. Lovelock (1976). "Continuous and Ultrasensitive Apparatus for the Measurement of Air-Borne Tracer Substances," *Journal of Chromatography*, Vol. 126, pp. 3–9.

Singh, G. P., et al. (1990). "TubeStat—An Innovative Approach for Maintaining Heat Exchanger History," EPRI NDE BOP Workshop, New Orleans, August.

Smith, A. (1993). "Reliability-Centered Maintenance," POWER-GEN Americas, Dallas, TX, November.

Sparrow, E. M., W. J. Minkowycz, and M. Saddy (1967). "Forced Convection Condensation in the Presence of Noncondensables and Interfacial Resistance," *Journal of Heat and Mass Transfer*, Vol. 10, pp. 1829–1845.

Spencer, R. C., K. C. Cotton, and C. N. Cannon (1974). "A Method for Predicting the Performance of Steam Turbo-Generators —16,500 kW or Larger," based on ASME Paper 62-WA-209, ASME Winter Annual Meeting, November 1962, revised July.

Stanton, D. J. (1995). "Economic Impact of Condition-Based Maintenance on Power Generation Cost," *Proceedings, POWER-GEN Europe 1995*, Amsterdam, Vol. 7, pp. 597–615.

Stiemsma, R. L., G. K. Bhayana, and R. D. Thurston (1994). "Performance Enhancement by an Innovative Tube Cleaning Application," *Proceedings of IJPGC 1994*, Phoenix, AZ, American Society of Mechanical Engineers, New York, PWR-25-1994, pp. 7–11.

Strauss, S. D (1992). "Advanced Tracer Technique Enhances Leak Detection," *Power*, October, pp. 112, 114, 116.

Straus, S. D. "Condenser Full-Length Coatings: Linings for Condenser Tube Repair," *Power*, January, pp. 61–62

Szklarska-Smialowska, Z. (1986). *Pitting Corrosion of Metals*, National Association of Corrosion Engineers, Houston, 77084.

Taborek, J., and J. L. Tsou. "Recent Development in Power Plant Condenser Design Practice," *Proceedings, Heat Transfer Conference*, San Diego, August, American Institute of Chemical Engineers, New York, pp. 228–240, A.I.Ch.E. Symposium Series 288.

Tanasawa, I. (1989). "Condensation Heat Transfer—Japanese Research in 1980's," *Trans. Japan Society of Mechanical Engineers*, pp. 377–395.

TEMA (1988). *Standards of the Tubular Exchanger Manufacturers Association*, 7th ed., Tarrytown, NY.

Tombaugh, R. S. (1989). "Alternative Methods for Manganese Removal," EPRI Service Water Systems Reliability Improvement, Charlotte, NC, November 6–8.

Trevors, J. T., and C. M. Cotter (1990). " Copper Toxicity and Uptake in Microorganisms," *Journal of Industrial Microbiology*, Vol. 6, pp. 77–84.

Tsou, J. L. (1994). "New Methods for Analyzing Condenser Performance," *Proceedings, 1994 EPRI Heat Rate Improvement Conference*, Baltimore, May.

Tsou, J. L. (1996). Private communication, February 13.

Tucci, M., and R. J. Bell (1999). "Returning the Craig Unit 3 Condenser to Leak Tight Operation," *Proceedings, EPRI Condenser Technology Conference*, Charleston, SC, September 1–3.

Ullman' s Encyclopedia of Industrial Chemistry, (1988) Ed. Wolfgang Herhartz, etc., Vol. A 11, p. 338.

Underwood, A. G. V. (1934). "The Calculation of the Mean Temperature Difference in Multi-Pass Heat Exchangers," *Journal of the Institute of Petroleum Technology*, Vol. 25, p. 145.

VanDoren, P. "The Costs of Reducing Carbon Emissions," Briefing Paper No. 44, Cato Institute, Washington, DC, March 11.

Wenzel, L. (1962). "Measurement of the Performance of Condenser Tubes," *Proceedings, American Power Conference*, Vol. XXIV, pp. 617–629.

Wilde, D. J. (1964). *Optimum Seeking Methods*, Prentice Hall, Englewood Cliffs, NJ, pp. 24–30.

Wolff, P. J., P. A. March, and H. S. Pearson (1996). "Using Condenser Performance Measurements to Optimize Condenser Cleaning," *EPRI 1996 Heat Rate Conference*, Dallas, TX, May 22–24.

INDEX